BAHAMAS PRIMARY

Mathematics Book 3

The authors and publishers would like to thank the following members of the Teachers' Panel, who have assisted in the planning, content and development of the books, led by Dr Joan Rolle, Senior Education Officer, Primary School Mathematics, Department of Education:

Deidre Cooper, Catholic Board of Education

Vernita Davis, Ministry of Education Examinations and Assessment Division

LeAnna T. Deveaux-Miller, T.G. Glover Professional Development and Research School

Joelynn Stubbs, C.W. Sawyer Primary School

Dyontalee Turnquest Rolle, Eva Hilton Primary School

Karen Morrison and Daphne Paizee

The Publishers would like to thank the following for permission to reproduce copyright material.

Photo credits

All photos © Mike van der Wolk, Tel: +27 83 2686000, mike@springhigh.co.za, except: page 33 (r) © BananaStock – Thinkstock; page 147 © TMAX – Fotolia

Orders: please contact Bookpoint Ltd, 130 Park Drive, Milton Park, Abingdon, Oxon OX14 4SE. Telephone: (44) 01235 827720. Fax: (44) 01235 400454. Email education@bookpoint.co.uk Lines are open from 9 a.m. to 5 p.m., Monday to Saturday, with a 24-hour message answering service. You can also order through our website: www.hoddereducation.com

ISBN: **9781 4 718 6459 9**

© Cloud Publishing Services 2016

First published in 2016 by
Hodder Education,
An Hachette UK Company
Carmelite House
50 Victoria Embankment
London EC4Y 0DZ

www.hoddereducation.com

Impression number   10 9 8 7 6 5 4 3

Year   2020

Cover photo © Giovanni Costa/123RF.com

Illustrations by Peter Lubach and Aptara

Typeset in India by Aptara Inc.

Printed in India

A catalogue record for this title is available from the British Library.

Contents

Topic 1 Getting Ready Workbook pages 1–2

▲ Count the stars. What is double this number? What is half this number?

Do you remember what you learned in Grade 2? This year, you are going to **count** to higher numbers and learn more ways of calculating and solving problems. Before you start, you are going to revise some of the things you should remember from last year.

Getting Started

1 There are five points on each star. How could you quickly count all the points?

2 What is 10 lots of five?

3 What pattern can you see in the stars?

4 What colour is the sixth star?

5 In which position is the yellow star?

6 What fraction of the stars are:

   a blue   b red?

7 How many more stars would you need to have:

   a 40   b 100   c 93?

Key Words
count
skip-count
digit
place value
shape
patterns
graph

Unit 1 Number Work

Let's Think ...

Nadia makes 6 groups of ten stickers and has 7 left over.

● How many does she have altogether?

● How many more does she need to have 100?

You can already **count** *to 999 by ones and* **skip-count** *in groups of 2, 5 and 10.*

You can use the **digits** *from 0 to 9 to make any number. The value of each digit depends on its place in the number (***place value***).*

1 What is the value of the blue digit in each number?

   a 122   b 454   c 876   d 908   e 113

2 Count from 180 to 220 by fives. Write the numbers you count. Circle the odd numbers.

3 What is ten more than: a 70   b 120   c 143   d 190   e 195?

4 What is ten less than: a 90   b 140   c 197   d 300   e 304?

5 Write **<**, **=** or **>** in the boxes to make each statement true.

   a 132 ☐ 100 + 30 + 2   b 342 ☐ 432   c 543 ☐ 345

6 You are given the digits 3, 6 and 9. What is the greatest and smallest 3-digit number you can make with these three digits?

Looking Back

Write a number that:

a is greater than 310 but smaller than 315

b is 100 more than 342.

Unit 2 A Mixed Bag

Let's Think …

● Marie finds an object in the kitchen. It has two round faces. It can stack, slide and roll. What shape could it be?

● Jermaine finds a shape that has six square faces. What shape has he found?

You learned about flat **shapes** and solid shapes last year. You also counted shapes in your environment, found shape **patterns** and drew **graphs** to show how many of each shape you found.

1 Look at this graph.

  a What does it show?

  b How many cubes were there?

  c How many spheres and cones were there altogether?

  d What is the difference between the number of rectangular prisms and the number of cones?

  e How many shapes were counted in all?

Shapes in the Classroom

Type of Shape

- Cubes
- Rectangular prisms
- Spheres
- Cones
- Pyramids

0   4   8   12   16   20
Number of Shapes

2 Find a shape pattern in your environment. Draw it and write a few sentences to describe it.

Looking Back

Nathaniel folded some shapes in half.
Draw what you think the whole shape could be.

Topic Review

What Did You Learn?

● We revised some of the things we learned last year.

Talking Mathematics

The clues for this crossword puzzle have been lost.

Work in pairs to make up a clue for each word.

		¹H				²F				
		A				I				
		L		⁴P		F		⁵C		
		³F	R	A	C	T	I	O	N	
				T		H		U		
				T				N		
		⁶C	U	B	E			⁷T	E	N
				R						
				N						

Quick Check

1 Find numbers in the box to answer the questions.

399	333	386	115	311	395	270	165

a Which is the greatest number?

b Which is the smallest number?

c Which odd number has 9 in the ones place?

d Which number has 3 in the tens place?

e Which numbers have a difference of 4?

f Which number is 50 more than the fourth number?

g Which of these numbers would you count if you skip-counted by 5s from 100 to 400?

h Which numbers are even?

2 Which solids are these students talking about?

It has no faces and it can roll but not stack.

It has six faces but the faces are not all the same shape or size

Topic 2 Numbers in Our World

Workbook pages 3–6

Meals just $6 each
Pizza, salad and water
Fish, salad and juice.
Burger and juice.

Key Words
counting
ordinal numbers
position
cardinal numbers
Roman numerals

▲ What would a family of four pay if they each ordered a meal at this restaurant? How did you work this out?

We use numbers all the time in our daily lives. Money, telephone numbers, licence plates and house addresses all use numbers. In this topic, you are going to learn more about **ordinal numbers**. You are also going to investigate **Roman numerals** and think more about how we use numbers in everyday life.

Getting Started

1 If you come sixth in a race, how many people finish in front of you?

2 What do the numerals XII mean? You will often see them on a clock.

3 Do you need to be able to add and subtract numbers when you go shopping? Why or why not?

4 Make a list of five different places in your community where you can see numbers.

Unit 1 Ordinal Numbers

Let's Think …

These athletes did well in a race
they entered. How do we describe this?

Shakima was ___.

Ashley was ___.

Sierra was ___.

Cardinal numbers *are the numbers that we use for* **counting***. They
tell us* how many *there are of something.* **Ordinal numbers***, such
as 1st, 2nd, 3rd and 4th, tell us about the* **position** *or order.*

Look at the ways in which we write ordinal numbers:

1st	first	11th	eleventh
2nd	second	21st	twenty-first
3rd	third	33rd	thirty-third
4th	fourth	40th	fortieth

1 Describe the position of each student in this race; for example:

The girl in the red sack is third.

2 Write the following numbers as ordinal numbers.

7   13   22   36   41   50

3 Work in pairs. Ask your partner questions about this shopping list. Use ordinal numbers; for example:

What is third on the list?

bread
apples
milk
ice cream
tomatoes
chicken
yoghurt
bananas

4 Read what these people are saying. Answer the questions.

It is my sister Kate's twenty-first birthday next month.

He is the seventeenth president of our club.

Misha came sixteenth out of a hundred in the drawing competition.

a How old is Kate this month?

b How many presidents did the club have before this person became president?

c How many people did better than Misha in the drawing competition? What was the position of the person who came last?

5 Ahkeem and Denison are going camping in the first week of July. They are going for 5 days. They leave on a Saturday, which is the second day of the month. On which dates are they going to be away?

Looking Back

a Write these numbers as ordinal numbers.

21   17   6   10   33   18   42

b Write your answers to part **a** in order.

Unit 2 Roman Numerals

Let's Think …

● Can you tell the time on this clock?

● Find the numbers that represent 12, 6, 9 and 3.

Roman numerals *were used in Ancient Rome more than 2000 years ago. The numbers are written using letters of the alphabet such as I, X and V. Each letter represents a number.*

I	1	VI	6	XI	11	XVI	16	XXI	21
II	2	VII	7	XII	12	XVII	17	XXII	22
III	3	VIII	8	XIII	13	XVIII	18	XXIII	23
IV	4	IX	9	XIV	14	XIX	19	XXIV	24
V	5	X	10	XV	15	XX	20	XXV	25

1 Write the answers to these questions in Roman numerals.

a How old are you?

b What is the date today?

c How many people are there in your family?

d How many players are there in a cricket team?

e Which class are you in this year?

f My cousin is 17 years old. Her best friend is one year older. How old is this?

g What is your telephone number?

2 Which letter of the alphabet represents each of these numbers?

a 1   b 5   c 10

> To make Roman numerals, you have to add or subtract the smaller letter values from the greatest one. If the smaller value letters are after the greatest one, you add them. If they are before it, you subtract them; for example:
>
> 8 = VIII (5 + 3)
> 9 = IX (10 – 1) (1 before 10)
> 12 = XII (10 + 2)

3 Write these numbers as Roman numerals.

a 3   b 8   c 15   d 19   e 27   f 22

4 Danny wrote his friend a message in a secret code. Can you work out what he said?

Please call

XII   XIII   I   IX   IV   II   V

5 Competitions like the Super Bowl use Roman numerals. Look at this logo for a sporting event.

a Can you read the Roman numerals on the logo? Hints: L = 50 and you have to subtract.

b What do you think this number means? Why is it included on the logo?

ALL STARS
KNOCKOUT
CHALLENGE
XLIX

Looking Back

Write down the numbers that match these Roman numerals.

a XIV   b XII   c IX   d XX   e IV   f XIII

Unit 3 Numbers We Use Every Day

Let's Think ...

Look at these photographs.
Why do we need to know about numbers
to use all of these things?

We use numbers all the time without really thinking about them. You are going to use some of your mathematics skills and knowledge to work with the sorts of numbers we find around us every day.

1 Add up the digits on each car licence plate as quickly as you can.

2 Put these temperatures in order, starting with the coolest.

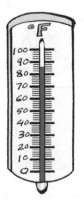
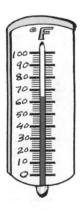
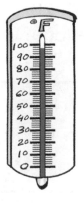
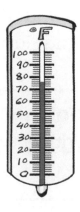

3 Read the advertisement for the T-shirts that are on sale.

a How much will you pay for one T-shirt?

b How much will you pay for two T-shirts?

c How much will you pay for four T-shirts?

4 Look at the measurements of this table. How much fabric would you need to make a tablecloth for the table? The fabric is 120 cm wide. You need to have 10 cm hanging over at each end.

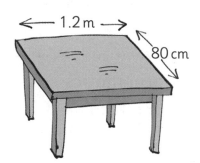

5 Read these street addresses. Put them in order. Then draw a diagram of the street to show where the houses are.

| 19 New Street | 21 New Street | 23 New Street |
| 20 New Street | 22a New Street | 22b New Street |

Looking Back

Make a list of 5–10 ways in which numbers are used every day. For each, explain why you need to understand numbers.

Topic Review

Talking Mathematics

What is the mathematical word for:

- numbers we use for counting how many of something there are
- numbers that show position
- numbers that use the letters of the alphabet?

Quick Check

1 Complete the sentences. Choose the correct number or word.

  a There were (35/35th) entries in the cycle race. Wallace came (21/21st).
    Ellis came (35th/last) because he had a puncture.

  b It is Standika's (8/8th) birthday today. Her friends will come to her party.
    Her friends are all (8th/8) years old.

2 Which of these telephone numbers do you think will add up to a greater number? Guess, and then work out the answer.

  a 678 259   b 217 230

3 Write these numbers as Roman numerals.

  a 7   b 12   c 16   d 20   e 19

4 Write down the numbers that match these Roman numerals.

  a IV   b VIII   c IX   d XIII   e XVII

Topic 3 Exploring Patterns

Workbook pages 7–9

Key Words

pattern
element
repeating pattern
growing pattern
number line
ascending
descending

▲ Look at the beads carefully. What patterns can you see? Try to find at least three different patterns. Describe the patterns you find to your partner.

Patterns are very important in mathematics. We use number patterns when we count, shape patterns to help solve problems involving shape and space, and repeating patterns to help us remember number facts and do mental calculations.

Getting Started

1 Look at the beads in the photograph again. Draw what you think the next part of the pattern will look like.

2 Can you make five different patterns using only the numerals 0 and 1? Write your patterns in your book.

3 How are these two patterns similar? How are they different?

Unit 1 Repeating and Growing Patterns

A **pattern** is something that follows a particular rule. You can see how the pattern works and you can predict what the next pattern **element** will be.

In a **repeating pattern**, the same elements are repeated to form the pattern.

In a **growing pattern**, the pattern elements grow with each step in the pattern.

In this growing shape pattern, two more squares are added to each element to make the pattern grow:

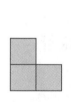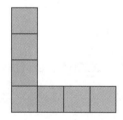

In this growing number pattern, the numbers go up in twos to make the pattern:

0, 2, 4, 6, 8, 10

In this growing pattern made with letters, one more x is added to each element to make the pattern grow: xy   xxy   xxxy   xxxxy

1 Work in pairs. Take turns to describe the patterns. Say whether the pattern is a repeating pattern or a growing pattern.

a	2	5	5	2	5	5	2		
b	2	5	2	2	5	2	2	2	5
c	3	13	23	33	43	53			
d	A	b	b	A	b	b	A	b	

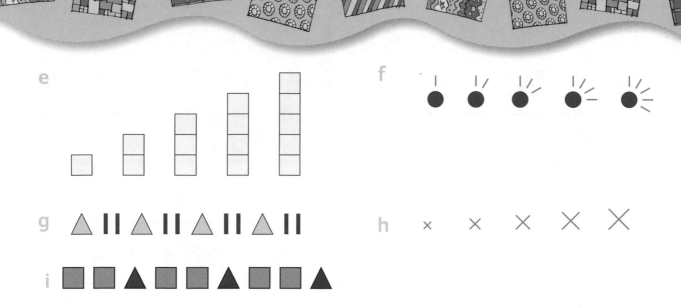

e

f

g △ ‖ △ ‖ △ ‖ △ ‖

h × × × × ×

i ■ ■ ▲ ■ ■ ▲ ■ ▲

2 Write or draw the next two elements for each of the patterns in question 1.

3 You are going to design patterns using potato or cork prints.

   a Follow the instructions in the pictures to make your printer.

   b Print your own repeating pattern.

   c Print your own growing pattern.

Instructions

Carve out your shape.

Dip the potato into paint.

Press onto paper to print.

Looking Back

1 Use a circle and a triangle to make:
  a a repeating pattern   **b** a growing pattern.

2 Swap patterns with a partner. Describe each other's patterns in words.

Unit 2 Patterns and Number Lines

Let's Think ...

- Work out the missing numbers in each pattern.
- Say how you found the missing numbers.
- Describe each pattern.

1 3 5 ✦ ✦ 11	5 10 ✦ ✦ 25 30
100 90 ✦ ✦ 60	63 ✦ ✦ 33 23

To find the missing numbers in a pattern, you have to decide whether the numbers are in **ascending** or **descending** order.

Ascending numbers go up in value; for example:

10   20   30   40

Descending numbers go down in value; for example:

100   95   90   85

You can use a **number line** to find the pattern and work out the missing numbers. Look at these examples carefully.

What is the missing number?

14   20   ☐   32

What is the missing number?

125   ☐   115   110

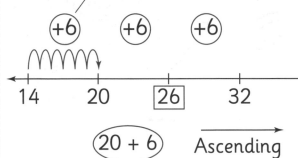

work out the difference

(+6) (+6) (+6)

14   20 26 32

(20 + 6)   Ascending →

The numbers are in ascending order so we count on or add.

work out the difference

(−5) (−5) (−5)

110 115 120 125

← Descending   (125 − 5)

The numbers are in descending order so we count back or subtract.

16

1 Work out the missing numbers. Then copy the patterns and write the next element in each one.

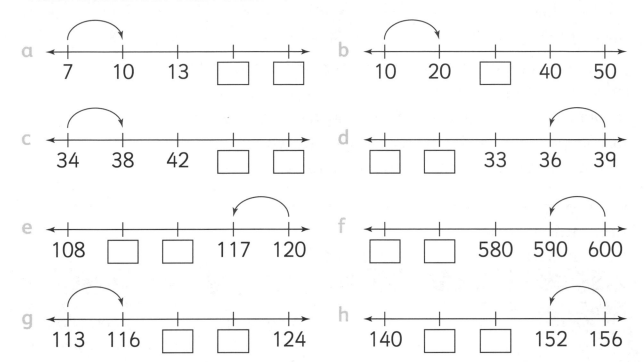

a 7   10   13   ☐   ☐

b 10   20   ☐   40   50

c 34   38   42   ☐   ☐

d ☐   ☐   33   36   39

e 108   ☐   ☐   117   120

f ☐   ☐   580   590   600

g 113   116   ☐   ☐   124

h 140   ☐   ☐   152   156

2 Follow the instructions and write the first five numbers in each pattern. Use a number line if you need to.

   a Start at 21. Add 6 each time.   b Start at 60. Count back by twos.

   c Start at 100. Count by tens.   d Start at 175. Subtract 10 each time.

   e Start at 34. Count by threes.   f Start at 163. Count back by tens.

3 Write a rule for each number pattern.

   a 30,   35,   40,   45,   50   b 179, 169, 159, 149, 139

   c 4,   8,   16,   32   d 48,   24,   12,   6

Looking Back

a Make up your own number pattern and leave out two numbers.

b Swap with a partner. Work out the missing numbers in each other's patterns.

Topic Review

Talking Mathematics

Match each pattern to the correct description.

105, 95, 85, 75	135, 140, 145, 150	AB AAB AAAB	ABA BAB ABA
A repeating pattern made using the letters A and B.	A growing pattern made using the letters A and B.	A number pattern in which the numbers increase by 5 each time.	A descending pattern where numbers go down by 10 each time.

Quick Check

1 Use the letters A, B and C to make: a a repeating pattern b a growing pattern.

2 Write the next three numbers in each pattern.

   a 45, 40, 35   b 127, 137, 147

3 Read this description and draw the pattern.

   > This repeating pattern is made by drawing triangles in rows, facing each other to make diamond shapes. There is a small square inside each diamond shape.

4 Use patterns to help you find the answers to these questions.

   a There are 50 toes. How many people are there?

   b A kitchen stool has three legs. If there are 27 legs, how many stools are there?

   c In a field of goats, there are 44 legs. How many goats is this?

Topic 4 Counting and Place Value Workbook pages 10–12

Key Words
digit
place
place value
value
ones
tens
hundreds
thousands
expanded form
compare

▲ What is the name of these coloured sprinkles? What do you think this means? What is the difference between a hundred and a thousand?

Last year, you counted, wrote and worked with numbers from 0 to 999. This year, you are going to count even higher. You will also learn how to work with **place value** in numbers with more than three **digits**.

Getting Started

1 Say these numbers.

   a 243   b 342   c 234   d 423

2 Sondra has five $1.00 bills, three $10.00 bills and six $100.00 bills. How much is this altogether?

3 A pharmacy sells these ear buds.

   a How many ear buds in five containers?

   b How many ear buds in ten containers?

Unit 1 Place Value

Let's Think ...

Jocelyn cut these shapes out of number charts and rubbed out some of the numbers.

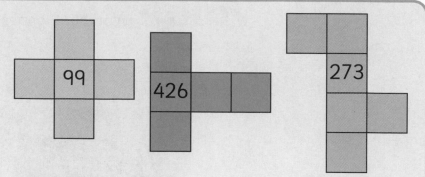

- Can you work out which numbers she rubbed out?

- How can you use number patterns to help you?

The number *328* has three **digits**.

We can show 3-digit numbers on a **place value** chart like this:

Hundreds	Tens	Ones
3	2	8

We can also write the number in **expanded form** like this:
328 = *300* + *20* + *8*.

The **place** of each digit tells us what **value** the digit has in the number. The *3* has a value of *3 hundreds* or *300* because it is in the **hundreds** place. The *2* has a value of *2 tens* or *20* because it is in the **tens** place. The *8* has a value of *8 ones* or *8* because it is in the **ones** place.

You can only write digits up to 9 in each place.
999 is the highest 3-digit number.
999 + 1 makes a thousand. We write *1 000*.
1 000 is a 4-digit number. It has four digits. Each digit has its own place value.
The place value table for 4-digit numbers looks like this:

Thousands	Hundreds	Tens	Ones
3	1	4	5

3 145 is three thousand one hundred forty-five.
We leave a space after the thousands digit to make it easier to read the numbers.
The *3* is in the **thousands** place, so it has a value of *3 thousands* or *3 000*.
We can write this in expanded form like this: *3 145* = *3 000* + *100* + *40* + *5*.

1 Say each number.

a 365   b 205   c 879   d 965

e 100   f 2400   g 3250   h 9000

i 2456   j 4050   k 3002   l 4098

2 Write these numbers using numerals.

a two thousand three hundred twenty-seven

b four thousand five hundred

c eight thousand three hundred forty-nine

d nine thousand nine hundred ninety-nine

e five thousand fifty

f five thousand five

3 What is the value of the 2 in each of these numbers?

a 235   b 325   c 532   d 523

e 2614   f 3215   g 5124   h 4132

4 Follow the instructions. Write the first five numbers that you count.

a Start at 3500. Count forwards by ones.

b Start at 4250. Count backwards by ones.

c Start at 900. Count forwards by hundreds.

d Start at 2400. Count backwards by hundreds.

e Start at 3459. Count forwards by tens.

f Start at 2564. Count backwards by tens.

5 Look at the numbers in the blocks.   3254   6024   4581   5124

Write the number that has:

a 4 ones and no hundreds

b 1 one and 5 hundreds

c 4 thousands and 5 hundreds

d 2 hundreds and 5 tens.

6 Use the digits 1, 3, 5 and 7 once only.

a What is the biggest number you can make?

b What is the smallest number you can make?

c How many different 4-digit numbers can you make using only these four digits?

> *Remember!* We use expanded notation to write a number as the sum of values in each place.
>
> *2 000 + 700 + 20 + 3 is the expanded notation for 2 723.*

7 Write each of these numbers as a sum using expanded notation.

a 24　　　　　b 98　　　　　c 129　　　　　d 450

e 5 465　　　　f 2 309　　　　g 3 120　　　　h 4 060

i 8 765　　　　j 9 990　　　　k 9 099　　　　l 5 498

8 Write these as ordinary numbers.

a 3 000 + 300 + 50 + 2　　　　　b 5 000 + 400 + 20 + 5

c 6 000 + 40 + 300　　　　　d 400 + 30 + 2 + 2 000

e 9 + 5 000 + 40　　　　　f 9 + 500 + 2 000 + 30

9 Write the missing value in each expanded notation.

a 2 345 = 2 000 + 300 + ☐ + 5　　　b 4 234 = ☐ + 200 + 30 + 4

c 9 876 = 9 000 + ☐ + 70 + 6　　　d 4 027 = 4 000 + ☐ + 7

e 3 006 = 3 000 + ☐　　　　f 4 060 = 4 000 + ☐

10 Say each number and then write it in words.

a 876　　　　　b 6 436　　　　c 2 090　　　　d 4 650

e 5 439　　　　f 4 999　　　　g 3 800　　　　h 1 909

Looking Back

Write the number shown on each of these place value mats.

a
Th	H	T	O
•	•	••	•
•		•	

b
Th	H	T	O
•	•		••
•			••

c
Th	H	T	O
••	•		••
••	•		••

Unit 2 Comparing Numbers

We **compare** numbers to find out which is greater and which is less.

500 is greater than 400   →   500 > 400.

2 500 is less than 2 550   →   2 500 < 2 550.

Remember, the symbol > means greater than and < means less than.
The open side of the symbol always faces the greater number.

You can compare numbers by placing them on a number line.
Numbers to the left are always smaller than numbers to their right.

You can also use a place value chart to compare numbers.
Start with the first pair of digits that are different, working from the left.

Look at these examples carefully.

Th	H	T	O
2	6	4	2
4	2	6	4

Th	H	T	O
2	6	4	2
2	6	2	4

2 thousands < 4 thousands
so, 2 642 < 4 264

4 tens > 2 tens
so, 2 642 > 2 624

23

1 Make ten pairs of numbers using the numbers in the box.
Write **<** or **>** to show which number is greater in each pair.

1 436	4 565	4 656	1 364	3 455	2 000
2 786	7 660	2 459	4 000	3 459	6 320

2 Sondra drew a number line marked in thousands.

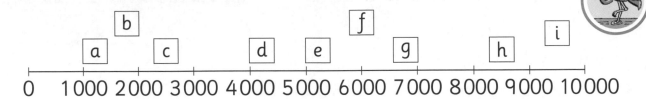

She ordered her numbers using the number line, but her number cards fell off!

a Work out where each card belongs on the number line.

b Write the numbers in ascending order.

c Write a number that would fit between card **a** and card **b**.

d Write a number for card **j**. It should be greater in value than card **i**.

3 Josh makes up a 4-digit password for his computer. It is the smallest number he can make with the digits 4, 6, 5 and 7. What is it?

4 Write each set of numbers in descending order.

a 1 988, 965, 1 898, 1 800   **b** 4 909, 6 709, 6 307, 4 037

5 Say whether each of these statements is true or false.

a 1 050 = 1 000 + 500   **b** 1 799 > 1 779

c 3 214 > 3 000 + 200 + 40 + 1   **d** 1 488 < 1 000 + 400 + 80

e 2 455 < 2 544   **f** 1 340 = 100 + 300 + 400

6 Jayson rolled four dice and got this result:

 a List the numbers he can make with a 1 in the thousands place.

 b What is the greatest number he can make?

 c Jayson makes the number 4 145. Write down the number that is:
  - one thousand less
  - ten more
  - 500 greater
  - two thousand greater.

7 Write the four numbers that come before each of these numbers.

  a 2 345   b 8 000   c 2 004   d 6 234

8 Arrange each set of money in order from least to greatest.

  a $4 125, $2 375, $3 872, $8 150, $4 324

  b $4 895, $1 199, $3 596, $4 250, $3 826

  c $5 412, $5 124, $5 421, $5 241, $5 142

9 Tori-Anne wrote these statements on the whiteboard. Marcus rubbed out some of the digits.

  1 2 5 0 > 1 ▮ 50   4 ▮ 19 < 4 1 2 1

  5 1 2 9 < 51 ▮ 9   6 9 9 9 > 5 9 8 ▮

  4 6 6 6 > 4 ▮ ▮ 6   4 1 9 9 > 4 ▮ 9 9

Work out what the missing digits could be.

Rewrite the statements so that each number has four digits.

Looking Back

1 Find the smallest number and the greatest number in each set.
2 Write a statement comparing the two numbers.

  a 1 249,   1 497,   1 963,   1 575

  b 2 672,   1 907,   1 364,   2 786

  c 4 743,   3 816,   4 629,   3 620

Topic Review

What Did You Learn?

- You can use place value to write any number using the digits from 0 to 9.
- A four-digit number has four place values: thousands, hundreds, tens and ones.
- Each digit has a different value which depends on its place in the number.
- You can use the symbols **<**, **=** and **>** to compare numbers.

Talking Mathematics

1 Read the clues. Write the numbers. Say each number in words.

  a I am a three-digit number between 600 and 700 with three tens and four ones.

  b I am a four-digit number with more than 8 thousands. I have the same number of tens and ones. I have three hundreds. The digit in the tens place has a value of 90.

2 Make up a clue of your own for a four-digit number. Swap with a partner and try to work out each other's numbers.

Quick Check

1 Write each number in figures.

  a two thousand three hundred seventy-seven

  b eight thousand four hundred eight

2 Write these numbers in expanded notation.

  a 6 568   b 8 056

3 What is 1000 less than:

  a 4 569   b 8 099   c 1 187?

4 What is the value of the 3 in each number?

  a 3 658   b 4 138   c 2 003

5 Write each set in ascending order.

  a 2 354,   2 453,   2 345,   2 435

  b 3 453,   4 212,   2 356,   3 089

6 Say whether each statement is true or false.

  a 3 459 < 3 954   b 8 560 > 8 660

7 Start at 4 889. Count by hundreds. Write the next five numbers.

Topic 5 Temperature
Workbook pages 13–14

▲ What are the children using to measure temperature?
Do you think it is hot or cold? Why?

Do you remember how to measure **temperature**? Now you are going to learn more about measuring temperature in two units of measure: degrees **Celsius** and degrees **Fahrenheit**. You will also learn how to **estimate** temperatures and then compare them with actual temperatures that you have recorded.

Getting Started

1 What is your body temperature when you are healthy?

2 What happens when a liquid gets very cold?

3 What happens when a liquid gets very hot?

4 Which scale do we normally use to measure temperature in The Bahamas: Celsius or Fahrenheit?

Key Words
temperature
Fahrenheit
Celsius
freeze
boil
thermometer
estimate

Unit 1 Reading Temperatures

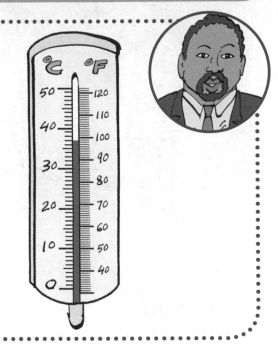

We use a **thermometer** to measure **temperature**. We can measure temperature in degrees **Celsius** or degrees **Fahrenheit**. In The Bahamas, we usually use the Fahrenheit scale.

We write temperatures like this:
37 °C
98.7 °F

The ° symbol means degrees.

1 Write down the temperatures shown on these thermometers in °C and °F.

a
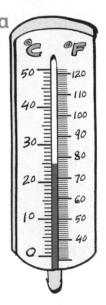

b
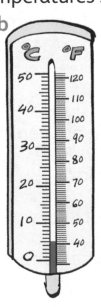

c
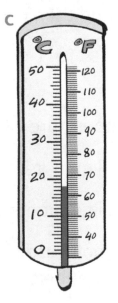

d
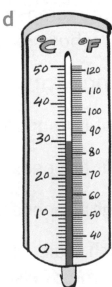

*Water **freezes** at 32 °F or 0 °C.*
*Water **boils** at 212 °F or 100 °C.*
A comfortable room temperature is 68 °F or 20 °C.

2 a Read and record the temperatures of the water in these pots.

  b In which pot is the water closest to boiling point?

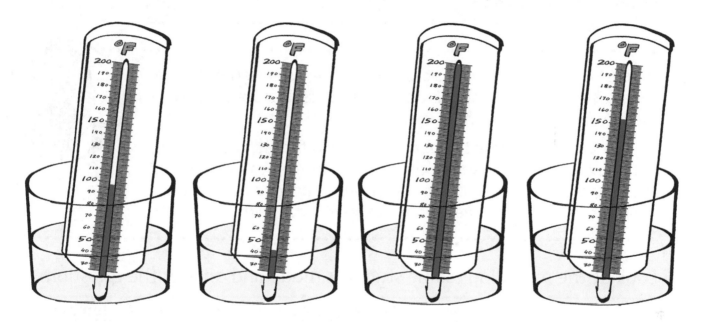

3 Imagine that it is very hot and you need to set the temperature on an air conditioner to make a room more comfortable. What temperature would you choose? Why?

Looking Back

Complete the sentences.

 a We use a ____ to measure temperature.

 b We can measure temperature in ____ Celsius or degrees ____.

Unit 2 Recording and Interpreting Temperatures

*Before we measure something accurately, we often **estimate** the measurement. An estimate is a sensible guess. You use what you already know to make the guess.*

If you have to estimate the temperature in a fridge, for example, you may estimate a low temperature such as 40 °F. You know that the temperature is low because it is cool inside a fridge, but you also know that the temperature must be above freezing point.

1 Look at the recordings some students made and answer the questions.

Place	Time	Our Estimate	Our Recording	Difference
Classroom	10:00 a.m.	70 °F	69 °F	1 °F
Under tree	10:10 a.m.	60 °F	58 °F	2 °F
Playground	10:20 a.m.	80 °F	92 °F	12 °F
School gate	10:40 a.m.	75 °F	97 °F	22 °F

a In which places did the students measure the temperature?

b At what time of day did they take their measurements?

c What was the highest temperature?

d What was the lowest temperature?

e Which temperatures did they estimate accurately?

f Which temperatures were different from their estimates?

2 Work in groups. You are going to estimate some temperatures and then measure actual temperatures.

a Draw a chart with columns or use the chart in your Workbook on page 14.

b Choose four places where you will measure the temperature.

c Discuss and estimate the temperature at each place.

d Measure and record the temperatures accurately.

e Work out the differences between your estimates and your recordings.

f Report back to the class.

3 Study this chart and answer the questions.

Day	Mon	Tues	Wed	Thurs	Fri
Place	outside office	outside office	outside office	outside office	outside office
Time	9:30 a.m.	9:30 a.m.	9:30 a.m.	9:30 a.m.	9:30 a.m.
Temperature	86 °F	84 °F	81 °F	84 °F	87 °F

a On which day was the temperature the highest?

b On which day was the temperature the lowest?

c On which days was the temperature the same?

4 a Work in pairs. Measure the temperature at the same place and at the same time every day for a week. Record the temperatures and report back to the class.

b Why should we measure the temperatures at the same time and place each day?

Looking Back

a What did you learn from Activity 2?

b Did you find any part of the activity difficult?

c What part did you enjoy the most?

Topic Review

Talking Mathematics

What is the mathematical word for each of these?

- a sensible guess based on what you already know
- something we use to measure temperature

Quick Check

1 Write down the temperatures shown on these thermometers.

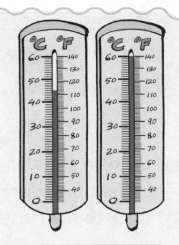

2 Study this chart and answer the questions.

Day	Mon	Tues	Wed	Thurs	Fri
Place	inside classroom	inside classroom	inside classroom	inside classroom	inside classroom
Time	9:30 a.m.	9:30 a.m.	9:30 a.m.	9:30 a.m.	9:30 a.m.
Temperature	56 °F	68 °F	77 °F	89 °F	104 °F

a On which day was the temperature the highest?

b On which day was the temperature the lowest?

c On which days was the temperature the most comfortable?

Topic 6 Talking About Time

Workbook pages 15–18

Key Words
minute
hour
day
week
month
year
calendar
a.m.
p.m.

▲ About how long does each of these activities take?
Which activity takes longer?

You already know that we talk about time in **minutes**, **hours**, **days**, **weeks**, **months** and **years**. In this unit, you are going to estimate and discuss how much time it takes to do certain activities. You will also learn about the difference between **a.m.** and **p.m.** times and do some more work with **calendars**.

Getting Started

1 How long does it take you to get to school in the morning?

2 What time of the day or night is 12 noon?

3 Which is longer, a minute or an hour?

4 Which is shorter, a week or a month?

5 Does your birthday fall on the same day of the week each year?

Unit 1 How Long Does It Take?

Let's Think …

Choose times from the pictures to answer these questions.

- At what time do you have breakfast?
- When do you watch television?
- At what time do you play?
- When do you get up in the morning?

*We measure time in **minutes**, **hours**, **days**, **weeks**, **months** and **years**.*

Some activities take a few minutes to complete, while others may take a few days or even years to complete.

1 Look at these activities. How long do you think each activity takes? Choose the best answer.

a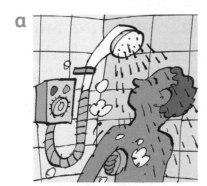

5 minutes
5 months
5 hours

b

4 days
4 hours
4 months

c

2 years
2 hours
2 minutes

d

1 minute
1 day
1 hour

2 Make a list of four things that you do every day. Then write down how much time you think each activity takes. Choose one of these descriptions for each activity.

- less than an hour
- more than an hour

3 Which of these activities will take more time?

a playing a game of cricket OR playing a game of football

b washing your face OR having a bath

c going to school OR going to visit your cousins on another island

4 Which of these activities will take less time?

a putting on your shoes OR getting dressed for school

b walking to school OR driving to school

c reading a whole book OR having breakfast

5 If it takes 10 minutes to have a shower every day and there are 7 days in a week, how long do you spend in the shower every week?

Looking Back
Complete these sentences.

a We measure time in minutes, ___, days, ___, ___ and years.

Some things take ___ time to do than others.

b It takes longer to play a game of cricket than it does to ___.

c It takes less time to have a bath than it does to ___.

35

Unit 2 A.M. or P.M.?

Let's Think …

- Is 4:00 p.m. in the morning or in the afternoon?
- When is 12 noon?
- Is 6:00 a.m. in the morning or the afternoon?
- When is 12 midnight?

We use **a.m.** as the abbreviation for the words ante meridian.
This means the time before noon or midday.
This time begins after midnight and continues until midday.

We use **p.m.** as the abbreviation for post meridian.
This means after noon or midday.
This time begins after midday and continues until midnight.

1 Choose a time to match each picture.

| 6:30 a.m. | 4:30 p.m. | 7:45 p.m. | 7:30 a.m. |

a

b

c

d

2 Work in groups. Make a list of 5 things that you do between 6:00 a.m. and 8:00 p.m. every day. Write down the times at which you do them.

3 Choose a time from the box to answer these questions.

  a You need to get to school by 8 o'clock in the morning. At what time should you leave home?

  b Your friend wants to go shopping on Saturday. At what time should you arrange to meet?

> 7:30 a.m.
>
> 7:30 p.m.
>
> 10:15 a.m.
>
> 10:15 p.m.

4 Kirkland wrote these times in his diary, but the times are not correct. Can you explain what the mistakes are?

Tuesday 5th September

12:00 a.m.

12:00 p.m.

Looking Back

a What does a.m. stand for?
b What does p.m. stand for?
c Is 4:00 p.m. after midday or after midnight?

Unit 3 Time

Let's Think ...

- How many minutes are there in an hour?
- Are there 12 or 24 hours in a day?
- What month is it now? How many days are there in this month?
- Which is shorter – a week or a month?

We measure time in different periods.

*An **hour** is a period of 60 **minutes**.*

*A **day** is a period of 24 hours.*

*A **week** is a period of 7 days.*

*A **month** is a period of 28, 30 or 31 days.*

*A **year** is a period of 12 months. There are 365 days in a year and 366 days in a leap year.*

Every 4th year is a leap year. February has 29 days in a leap year.

1 Work out the answers to these questions.

a How many days are there in two weeks?

b How many months are there in half a year?

c How many hours are there in 3 days?

d If a game lasts for 2 hours, how many minutes do the players play?

e If you sleep for 8 hours a day, for how many hours are you awake?

2 Complete these sentences.

    a There are 60 ___ in an hour.

    b There are 24 ___ in a day.

    c Some months have 30 ___ while others have 31.
       February has 28 days and ___ days in
       a leap year.

    d There are 12 ___ in a year.

3 Work in pairs. Use a clock and a calendar to show your partner how long these times are.

    a five hours   **b** a whole day   **c** three weeks   **d** two months

4 Torianne's birthday is on 25th October.
At what time does her birthday start?
When does it end?

5 Why is this statement incorrect?

A day is shorter than 60 minutes.

Looking Back

Say if these sentences are true or false. Correct the false sentences.

a A week is longer than 24 hours.

b There are seven months in a year.

c A month is shorter than a year.

d There are 45 hours in two days.

e An hour lasts for 60 minutes.

Unit 4 Using A Calendar

Let's Think …

Independence day in The Bahamas is always on the 10th of July. Is this always on the same day of the week? Look at two old calendars and find out.

A **calendar** *is a system we use for dividing years into months, days and weeks. A calendar shows dates, days and months. The dates in each month are not on the same day each year.*

In The Bahamas, we use the Gregorian calendar. There is also an Islamic calendar and a Chinese calendar.

1 Copy and complete this list of the names of the months of the year.

   January, ——, March, ——, May, June, ——,

   August, ——, ——, November, ——

2 a Which is the 2nd month of the year?

   b Which is the 4th month of the year?

   c Which month comes just after September?

   d Which month comes just before May?

3 Read the calendar and answer the questions.

   a What day of the week is the first day in October?

   b How many days are there in October?

   c How many Mondays are there in October?

   d Which day of the week is the 14th of October?

   e Which day of the week is the 28th of October?

			OCTOBER			
M	T	W	T	F	S	S
					1	2
3	4	5	6	7	8	9
10	11	★12	13	14	15	16
17	18	19	20	21	22	23
24	25	26	27	28	29	30
31						

f What is the date of the first Tuesday in October?

g On what day of the week does November begin?

h The 12th of October is a holiday in The Bahamas. Do you know what holiday this is?

4 Look at this year's calendar and answer these questions.

a On which day of the week is Emancipation Day?

b On which day of the week is Independence Day (10th July)?

c On which day of the week is your birthday?

d On which day of the week will people celebrate Christmas?

5 Imagine that you have to plan a birthday party on the 30th of October. Look at the calendar for October in question 3.

a You need to send out invitations at least two weeks before the party. By which date should you send out the invitations?

b You need to buy food and drinks two days before the party. When should you do this?

Looking Back

Complete the sentences.

a A calendar shows us dates, ___ and ___.

b The ___ month of the year is April and the 11th month is ___.

c There are ___ days in October every year.

d Independence Day is always on the same date, 10th of July, but it is not always on the same ___ of the week.

Topic Review

What Did You Learn?

- Activities take different amounts of time to complete.
- a.m. stands for *ante meridian*: the time before noon or midday.
  p.m. stands for *post meridian*: the time after noon or midday.
- We measure time in different periods: minute, hour (60 minutes), day (24 hours), week (7 days), month (28, 30 or 31 days), year (12 months).
- A calendar divides years into months, days and weeks.
- A date will not fall on the same day of the week each year.

Talking Mathematics

What is a mathematical word for each of these?
- a period of 12 months
- the time after midnight that continues until midday
- a system we use for counting years and for dividing years into days and weeks
- a period of 7 days

Quick Check

1 Copy and complete.

   a 7 days = ___   b 24 hours = ___   c 12 months = ___   d 60 minutes = ___

2 a.m. means ___. It is the time after ___ and before ___.

3 p.m. means ___. It is the time after ___ and before ___.

4 Answer the questions using the calendar for June.

   a How many Sundays are there in June?

   b What day of the week is the 23rd of June?

   c What is the date of the first Monday in June?

   d Is June shorter or longer than October?

JUNE						
M	T	W	T	F	S	S
		1	2	3	4	5
6	7	8	9	10	11	12
13	14	15	16	17	18	19
20	21	22	23	24	25	26
27	28	29	30			

Topic 7 Number Facts Workbook pages 19–21

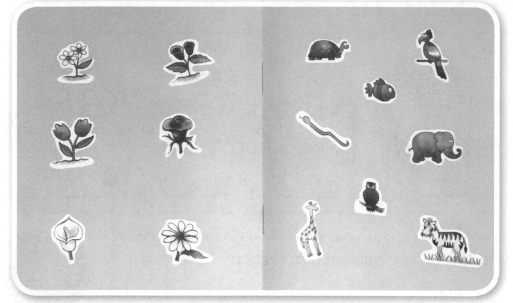

Key Words
add
subtract
addition
subtraction
fact
fact family

▲ How many stickers on each page? How many in total?
How many more are needed to make 20?

To answer the questions about the stickers, you could use the **addition** and **subtraction facts** that you already know. These facts are very useful in all areas of mathematics, so it is important to know them well and to be able to work them out mentally as quickly as possible. In this topic, you will practice working out number facts to 20 and revise some of the three-number **fact families** you worked with in Grades 1 and 2.

Getting Started

1 Think of any number from 1 to 20. Double the number you thought of. Then add 6. Halve the answer and subtract the number you started with. You should get 3.

2 Try this again with a few other numbers. Can you explain why you always get 3? Will this work if you start with 0?

3 How many ways can you find to get an answer of 15 by adding or subtracting numbers from 0 to 20? List all the ways you can find.

Unit 1 Addition and Subtraction Facts to 10

Let's Think ...

Each of the boxes should have 10 crayons. How many are missing from each box?

You should know the **addition** and **subtraction facts** to 10 very well by now.
You will use these facts when you do calculations with larger numbers.

1 Work out the missing score on each of these score boards.

a Total : 10
A	B
7	?

b Total : 8
A	B
4	?

c Total : ?
A	B
7	1

d Total : ?
A	B
5	4

e Total : 7
A	B
4	?

f Total : 10
A	B
0	?

g Total : ?
A	B
0	9

h Total : ?
A	B
1	8

i Total : 10
A	B
4	?

j Total : ?
A	B
5	5

k Total : 8
A	B
0	?

l Total : 9
A	B
3	?

Looking Back

a How many ways can you make 10 by adding two numbers? List them.

b Write three subtraction sentences with an answer of 5.

Unit 2 Addition and Subtraction Facts to 20

Let's Think ...

How do you know whether to add or subtract to find an answer?

There are different ways of working out the answers to addition and subtraction facts.

You should know all of the facts to 10 and all doubles from 1 to 10 by memory.

You can use these facts to help you work out other facts. Here are some useful strategies for getting an answer quickly.

Counting On and Counting Back

You can count on or count back to **add** or **subtract** small numbers (1 to 4).

2 + 17 = ☐   *Count on from the larger number: 17, 18, 19.*

19 − 1 = ☐   *Count back to subtract a small number: 19, 18.*

Doubles and Near Doubles

Doubles are useful for addition facts.

7 + 8 = ☐   *Double 7 is 14. One more is 15.*

Making Tens

You can work in steps and break up numbers to make 10.

9 + 7 = ☐   *Take 1 from the 7 and add it to the 9 to make 10.*
                     Then add 6 more. 10 + 6 = 16

17 − 9 = ☐   *Take away 7 from 17 to make 10. Then take away 2 more.*
                     10 − 2 = 8

Use Jotting to Make Jumps on a Blank Number Line

Draw a blank number line in your working space and jump in smaller steps.

8 + 9 = ☐   20 – 12 = ☐

8 + 9 = 17

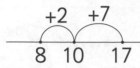

20 – 12 = 8

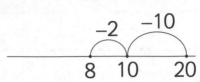

Fact Families

A **fact family** *is a set of related addition and subtraction facts using the same three numbers.*

13 + 7 = 20   7 + 13 = 20   20 – 7 = 13   20 – 13 = 7

If you know one fact, you can use it to work out the others.

1 Add. Use the strategy that you find easiest for each sum.

   a 11 + 7 = ☐   b 12 + 6 = ☐   c 9 + 2 = ☐

   d 8 + 7 = ☐   e 9 + 8 = ☐   f 16 + 3 = ☐

   g 15 + 5 = ☐   h 2 + 17 = ☐   i 9 + 11 = ☐

2 Subtract. Use the strategy that you find easiest for each subtraction.

   a 20 – 4 = ☐   b 18 – 9 = ☐   c 17 – 10 = ☐

   d 20 – 13 = ☐   e 17 – 8 = ☐   f 19 – 11 = ☐

   g 17 – 0 = ☐   h 12 – 9 = ☐   i 14 – 8 = ☐

3 Work out the missing numbers as quickly as you can.

a 16 + ☐ = 19   b 11 + ☐ = 14   c 12 + ☐ = 20

d ☐ + 3 = 19   e ☐ + 11 = 20   f 7 + ☐ = 19

g 20 − ☐ = 8   h ☐ − 4 = 15   i 20 − ☐ = 10

4 One fact in a family is given below. Write the three other facts.

a 12 + 8 = 20   b 11 + 5 = 16   c 20 − 7 = 13

d 9 + 8 = 17   e 18 − 3 = 15   f 5 + 14 = 19

g 18 − 8 = 10   h 17 − 9 = 8   i 16 − 7 = 9

5 Shandra has 8 red beads and 7 blue beads. Nicole has 7 red beads and 8 blue ones. Which girl has more beads?

6 A T-shirt costs $15.00.

a Brad has $13.00. How much more does he need?

b Jayson needs $8.00 more to buy a T-shirt. How much does he have already?

c Shawn has a $10.00 bill and a $5.00 bill. How much more does he need?

d Kevin has $20.00. How much change will he get if he pays for a T-shirt?

Looking Back

Use the three numbers in each set to write four different number sentences. Each number sentence must use all three of the numbers.

a 15,   1,   14

b 17,   4,   13

c 8,   9,   17

Unit 3 More Adding and Subtracting

Let's Think ...

- Add. Say what you notice.

    $12 + 0 = \square$ $\qquad$ $18 + 0 = \square$ $\qquad$ $17 + 0 = \square$

- Subtract. Say what you notice.

    $12 - 12 = \square$ $\qquad$ $19 - 19 = \square$ $\qquad$ $20 - 20 = \square$

$12 + 0 = 12$	Adding 0 does not change the value.
$12 - 0 = 12$	Subtracting 0 does not change the value.
$12 - 12 = 0$	Subtracting a number from itself leaves you with 0.
$12 + 6 = 6 + 12$	You can add in any order and you will get the same answer.
$12 - 3 \neq 3 - 12$	You cannot change the order in subtraction.

1 Decide whether these statements are true or false.

    a $12 + 7 = 7 + 12$ **b** $12 - 7 = 7 - 12$ **c** $19 + 0 = 0$

    d $19 - 0 = 0$ **e** $20 - 0 = 20$ **f** $17 - 2 = 2 - 17$

    g $11 + 8 = 19 + 0$ **h** $12 - 3 = 9 - 0$

2 Sammy needs $20.00. He only has $3.00. How much more does he need?

3 Torianne has 19 crayons. She loses her pencil case with all the crayons in it. How many does she have left?

Looking Back

Write three different number sentences to make each amount.

a $19.00   **b** $20.00   **c** $0.00

Topic Review

Talking Mathematics

Match words or phrases that have the same meaning.

take away	subtract	minus	plus	add
find the sum	find the difference	total	sum	answer

Quick Check

1 What are the missing numbers in this pattern?

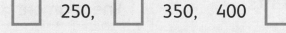

☐ 250, ☐ 350, 400 ☐

2 two thousand, nine hundred seven
  a Write this number in figures.
  b Is it odd or even?
  c Write it as a sum in expanded notation.

3 What is the value of 3 in each of these numbers?
  a 2340   b 3240
  c 2430   d 2043

e Write the numbers in parts **a** to **d** in ascending order.

4 Write the missing numbers.
  a 7 + 3 = ☐

  b 10 − 3 = ☐

  c 13 + ☐ = 20

5 Write the fact family for 13 + 5 = 18.

6 True or false?
  a 18 − 9 = 9 − 18
  b 9 + 7 = 7 + 9
  c 18 − 0 = 18 + 0
  d 20 − 0 = 0

Topic 8 Classifying Shapes

▲ What shapes can you see in the photograph? Which shapes are flat and which are solid?

You can see shapes in many different places – from the food we eat to the equipment we use. Some shapes, such as **triangles**, have **sides** and **corners**. Other shapes, such as **circles**, are curved. In this topic, you are going to learn more about the properties of shapes and how to describe what shapes look like. You will also learn about **points**, **lines** and **line segments**.

Key Words
solid
plane
square
triangle
rectangle
circle
oval
sphere
cone
cylinder
cube
face
corners
sides
edges
closed shape
dimensions
point
line
line segment

Getting Started

1 Think of a shape that has three sides and three corners. What is it called?

2 Does a circle have corners?

3 Think of two road signs you have seen. Draw them. What shapes are they?

4 What shape is the face of your watch? Are all watch faces the same shape?

5 What can you use to draw a straight line?

Unit 1 Plane Shapes

Let's Think ...

- Which of these shapes have equal sides (sides that are the same)?
- Which shapes have four corners?
- Which shapes have no straight sides?
- Can a shape have two sides?

 circle

 triangle

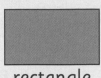 rectangle

 square

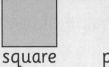 pentagon

 oval

Shapes like the ones in the box above are called flat shapes. We say that they lie on a flat surface or plane, which is why they are called **plane** shapes. Flat shapes have two dimensions: length and width.

Some plane shapes have straight parts, called **sides**. They also have **corners**, which is where the sides meet. They are **closed shapes** because all the sides meet.

1 Look at the shapes and answer the questions.

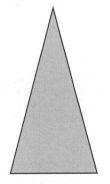

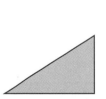

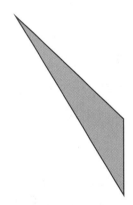

   a How many sides does each shape have?

   b How many corners does each shape have?

   c What do we call these shapes?

   d How are the shapes different?

   e How are the shapes similar?

Circles *and* **ovals** *are closed plane shapes with no straight sides and no corners. An oval looks like a circle that has been stretched.*

2 Put your finger on the middle point of this circle. Then measure the distance from the middle to anywhere on the outside of the circle. What do you notice?

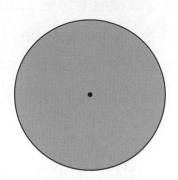

3 Look at these two shapes.

a How many sides does each shape have?

b How many corners does each shape have?

c What do we call A?

d What do we call B?

e What is the difference between the shapes?

A

B

4 What shapes can you make if you fold this shape in half?

Looking Back

Complete the sentences.

a Circles and ovals are ___ shapes. They do not have ___ or ___.

b A square has ___ sides. All the sides are ___ in length. A rectangle also has ___ sides, but the ___ are not all the same length.

Unit 2 Solid Shapes

Solid *shapes are shapes that have three* **dimensions***: height, length and thickness or depth.*

The flat parts of a solid shape are called **faces***.*

Some solid shapes have **edges** *and* **corners***. Some solid shapes are round.*

rectangular prism

sphere

cone

cylinder

pyramid

cube

1 Look at these everyday objects. What shapes are they?

a

b

c
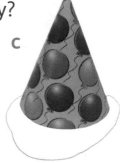

d

e

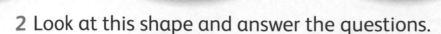

2 Look at this shape and answer the questions.

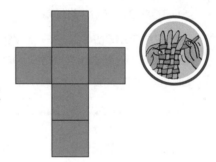

corner

edge

face

   a What type of shape is this? Is it a solid shape or a plane shape? How do you know?

   b How many faces does this shape have? Can you see them all?

   c How many corners does this shape have?

   d How many edges does it have?

3 Think about a cylinder.

   a How many faces does it have?

   b What is the shape of the faces?

   c How many corners does a cylinder have?

4 Look at this diagram.

   a What shapes can you see in the diagram?

   b If you cut the diagram out and fold it up, what type of shape would you be able to make?

   c Draw a diagram like the one above and make the shape. Describe the shape.

5 If you had to design a container for holding pencils, which shape would you choose for the design? Why?

6 Why is a cone a good shape for holding ice cream? Are there any other suitable shapes?

> ## Looking Back
> Draw the following shapes and write a sentence to describe each shape.
> **a** a cylinder   **b** a cube   **c** a sphere   **d** a cone

Unit 3 Lines, Points and Line Segments

*A **line** is straight. It goes in both directions and it does not end.*

*A **point** is a small dot that marks an exact place or position. A and B are points on this line.*

A ←————————→ B

*A **line segment** is any part of a line between two points.*

1 How many line segments can you draw between these points? Copy the points on a sheet of paper. Draw the segments and see if you are correct.

a    b    c    d

2 Draw and label the following line segments.

   a 6 cm   b 11 cm   c 2 in   d 8 in

3 Draw a circle. Mark five points on the circle. Use a ruler to join each point to the four other points with a line segment.

The picture shows you how to get started.

Compare your drawing with a partner.
How are the drawings the same? How are
they different?

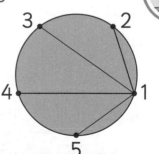

Topic Review

Talking Mathematics

What is the mathematical word for each of these?

- a shape with no thickness
- a flat round closed shape with no corners
- a shape with three dimensions

- the flat part of a solid shape
- the exact position or place
- a part of a line between two points

Quick Check

1 A square has ☐ sides and ☐ corners.

2 A triangle has ☐ sides. The sides (do/do not) have to be the same length.

3 A cube has ☐ faces, ☐ corners and ☐ edges.

4 A sphere is a solid, ___ shape.

5 The part of a line between two points is called a line ___.

Topic 9 Rounding and Estimating
Workbook pages 26–28

▲ This is a piece of a hand-knitted sweater. How many stitches do you think there are in this piece of knitting? How can the small square help you estimate?

There are many situations in real life where we do not need to know the **exact** number of items; for example, you may hear that about 5 000 people attended a cricket match. This means that there were around 5 000 people, but the actual number could be more or less than that. **Estimating** values and **rounding** numbers is useful in mathematics too. We estimate before we measure and we can estimate before we do calculations to get a rough idea of what the answer should be.

Getting Started

1 How many ants do you think there are on this sandwich? Choose one estimate.

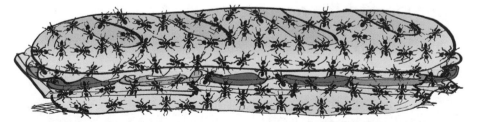

about 10

about 50

about 100

about 1 000

2 Find a quick way to count the ants. Write down how many you counted.

3 Check your answer with a partner. Tell each other how you counted the ants.

Unit 1 Rounding Numbers

Let's Think ...

Micah and Jermaine pick numbers without looking.
If the number is closer to 10, they get 10 points.
If the number is closer to 20, they get 20 points.

- How much would each boy score?
- How did you decide?
- What score would you give for 15? Why?

Micah picks 13, 16 and 9.

Jermaine picks 19, 16 and 11.

Place value *is helpful for rounding numbers.*

*Follow these steps to **round** a number to the **nearest** ten.*

- *Look at the **digit** in the ones place.*
- *If the digit is 0, 1, 2, 3 or 4, you leave the tens unchanged and write 0 in the ones place.*
- *If the digit is 5, 6, 7, 8 or 9, you round up to the next ten and write 0 in the ones place.*

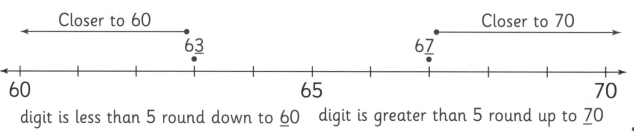

Closer to 60   6<u>3</u>   6<u>7</u>   Closer to 70

60   65   70

digit is less than 5 round down to <u>6</u>0   digit is greater than 5 round up to <u>7</u>0

1 Round each of these numbers to the nearest ten.

   a 19   b 52   c 58   d 72   e 67

   f 65   g 38   h 145   i 328   j 409

2 Which of these numbers will round to 50?

48	51	55	45	51	61	59	42	53

3 Veronique says that 180 people attended a church picnic. If she rounded the number of people to the nearest 10, list the actual numbers of people who could have attended.

We look at the tens digit to round numbers to the nearest hundred.

- *If the digit in the tens place is less than 5, the hundreds digit remains unchanged and you write 0 in the tens and ones places.*
- *If the digit in the tens place is 5 or greater, round up to the next hundred and write 0 in the tens and ones places.*

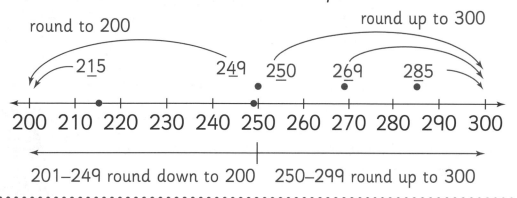

4 Round these numbers to the nearest hundred.

a 189   b 494   c 629   d 164   e 545

f 1 230   g 2 099   h 7 098   i 4 549   j 3 612

5 There were 2 807 spectators at a cricket match. A newspaper reporter rounded the number to the nearest ten and the ticket seller rounded the number to the nearest hundred.

a What number did the newspaper reporter use?

b What number did the ticket seller use?

6 Collect ten 3- or 4-digit numbers from newspapers or magazines. Stick them on a sheet and round each one to the nearest ten and hundred.

Looking Back

Say whether each of these statements is true or false.

a 426 rounds to 420   b 455 rounds to 460   c 1 245 rounds to 1 200

Unit 2 Estimating Answers

Let's Think ...

The numbers of visitors to an art gallery on four days were:

55,   48,   73,   51

Three children estimated the total number of visitors.

> This is about 100.

> I guess 180.

> I estimate 210.

Which estimate do you think is best? Why?

*If you do not need to know the **exact** answer, you can use rounded numbers to find an **approximate** answer.*

*The examples show how to use rounding to **estimate** the answers.*

83 – 47 = ☐

80 – 50	*Round the numbers to the nearest ten.*
The answer will be about 30.	*Think 8 tens – 5 tens = 3 tens.*

163 + 296 = ☐

200 + 300	*Round the numbers to the nearest hundred.*
The answer will be close to 500.	*Think 2 hundreds + 3 hundreds = 5 hundreds.*

1 Use rounded numbers to estimate the answer to each calculation.

a 19 + 32	b 58 + 37	c 42 + 82	d 49 + 23
e 91 + 56	f 92 – 42	g 75 – 38	h 76 – 43

2 Four brothers are aged 26, 22, 19 and 14. Estimate their total age.

3 There are 88 animals on a farm. 39 are goats. The rest are chickens. Estimate the number of chickens.

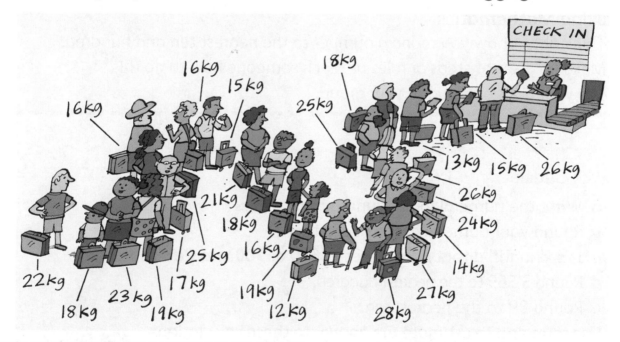

4 Estimate the answers to these calculations.

a 176 + 139   b 452 + 271   c 138 + 366

d 163 + 196   e 237 + 458   f 812 + 199

g 654 − 243   h 278 − 127   i 475 − 237

5 Work with a partner.

a Estimate how many kilograms of luggage these passengers have in total.

b The airline allows for 20 kilograms per person. How much would they allow for these passengers?

c According to your estimate, will there be too much luggage?

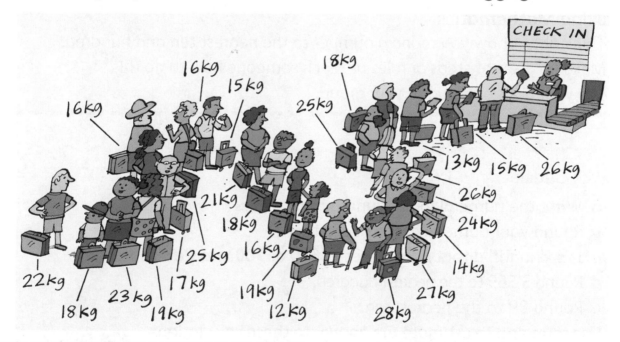

Looking Back

The numbers of students in different grades in a school district are given in the table.

Grade	1	2	3	4	5
Number of Students	286	312	363	319	391

a Use rounding to estimate the total number of students in grades 1 and 2.

b Estimate how many students there are in grades 1 to 5.

Topic Review

What Did You Learn?

- Place value is used to round numbers to the nearest ten or hundred.
- Use the digit in the ones place when you round to ten. If the digit is 5 or more, round up to the next ten. If it is less than 5, the ten remains unchanged.
- Use the digit in the tens place to round to the nearest hundred. Five or more tens means that you round up to the next hundred.
- An estimate is a careful guess. Rounding is useful for estimating.
- Round numbers to the nearest ten or hundred to estimate the answer.

Talking Mathematics

- Think about how you round a number to the nearest ten and hundred.
- Make up a set of steps or rules to teach someone how to do this.
- Say your rules out loud to your group.

Quick Check

1 a Write the numeral for six hundred thirty-five.

   b Round your numeral to the nearest ten.

   c Is six hundred thirty-five closer to 600 or 700? How do you know?

   d Round 3 569 to the nearest hundred.

   e Round 89 to the nearest ten.

2 Use rounding to estimate the answer to these calculations.

   a 65 + 18   b 174 − 32

3 Tenille used a calculator to do some calculations. These are her answers.

   326 + 550 = 776   743 − 457 = 186

   Without doing the actual calculations, how can you show Tenille that her answers are not correct?

4 Mrs Simpson wants to fence two lengths of a field. One length is 312 m and the other is 458 m. She buys 700 m of fencing. Will that be enough? Explain why or why not.

Topic 10 Number Patterns and Relationships

Workbook pages 29–32

Key Words
skip-count
interval
groups
forwards
backwards
pattern
odd
even

▲ Count the markers in each pack. How many are there altogether? Can you find the total without counting each marker? How?

You already know how to count and you should be able to count **forwards** and **backwards** from any number your teacher gives you. Counting in **groups** can help you to find a total quickly; for example, if you have seven dimes, you can work out how much money this is by counting in tens: 10, 20, 30, 40, 50, 60, 70. In this topic, you are going to practice counting by different numbers. You are also going to revise **odd** and **even** numbers and learn some of the rules that they follow.

Getting Started

1 Which numbers are missing from these number lines?

a   27   ?   ?   36   39   ?   45

Tell your partner how you worked out the missing numbers.

b   ?   ?   75   80   ?   ?   95

2 Mrs Guthrie has nine $10.00 bills. How much is this altogether?

Unit 1 Odd and Even Numbers

Let's Think …

There are 11 players in a team. They put their socks into the laundry when they have finished playing.

● Shareen counts 21 socks. Is that correct?

● Explain why or why not.

An **even** number of counters can be grouped into sets of two with none left over.

When an **odd** number of counters are grouped into sets of two, there is always one left over.

16 ⟶ even   17 ⟶ odd

odd one out

All even numbers have a 0, 2, 4, 6 or 8 in the ones place.

All *odd* numbers have a 1, 3, 5, 7 or 9 in the ones place.

1 Work in pairs. Find the path through this number grid. You may only move onto odd-numbered blocks that touch each other.

Start ↓

22	17	31	18	32	16	31	45	60	99
18	44	21	100	93	42	66	11	75	22
51	32	59	32	33	34	142	50	60	37
81	30	23	70	11	130	131	14	48	160
65	121	31	54	129	44	40	133	31	59
43	12	22	26	57	29	97	97	74	11
81	120	88	130	45	40	80	34	100	33
79	52	64	12	11	52	71	88	141	111
59	55	51	35	39	31	24	90	53	98
90	80	100	44	20	123	90	67	99	102

Finish ↑

2 Write the next three even numbers each time.

   a 12, 14, 16   **b** 34, 36, 38   **c** 102, 104, 106   **d** 400, 402, 404

3 Sort the numbers in the box into sets of odd and even numbers. Make two lists in your book.

14	43	57	68	99	100	32	108	17
199	800	422	132	950	906	207	321	809

4 Work with your group.

   a Choose any two even numbers. Add them. Is your answer odd or even?

   b Try adding some more even numbers. What do you notice?

   c Choose any two odd numbers. Add them. Is your answer odd or even?

   d Try adding some more odd numbers. What do you notice?

   e Now try adding one odd and one even number. What do you notice?

   f Complete these rules:

     even + even = ___ even + odd = ___ odd + odd = ___ odd + even = ___

     even − even = ___ even − odd = ___ odd − odd = ___ odd − even = ___

5 Zara is thinking of a number. It is an odd number. It is less than half of 14 and it is greater than 3. What is the number?

6 There are an even number of children in a classroom. The digits of the number add up to 3. How many children could there be?

Looking Back

1 List the even numbers from 30 to 50.
2 List the odd numbers from 71 to 51.
3 List the odd numbers you can make using the digits 1, 2 and 3.

Unit 2 Skip-counting

Let's Think …

Andrew counted out exactly
40 pencils in equal groups.

● What groups could he have counted in?

● Try to think of a group that he could not
have used. Explain why it would not work.

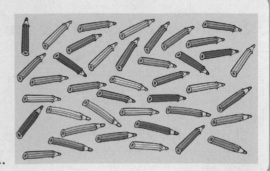

When we count by **groups**, we skip some numbers; for example, if we
count by 2s from 0 to 10, we count 0, 2, 4, 6, 8, 10. When we count
by 2s, the counting **interval** is 2. That means we only count every
second number and skip the numbers in between.

When we count by 5s, the interval is 5 and we only count every fifth number.
You can **skip-count** both **forwards** and **backwards**.

1 Keishla skip-counted by 2s
and coloured yellow the
numbers she counted. Then
she skip-counted by 4s and
circled the numbers she counted.

1	2	3	④	5	6	7	⑧	9	10
11	⑫	13	14	15	⑯	17	18	19	⑳
21	22	23	㉔	25	26	27	㉘	29	30
31	㉜	33	34	35	㊱	37	38	39	㊵

a What patterns can you see in the chart?

b When you count by 4s, which numbers do you count?

c Why did you not count any odd numbers?

2 Skip-count by 2s from 66. What is the next number that you count with
the same digits in the tens and ones places?

3 The winning ticket number in a raffle is between 60 and 80.
You would count this number if you counted by 2s and if you
counted by 4s. The number has one odd and one even digit. The
difference between the digits is 5. What is the winning number?

4 Skip-count in the questions below. Write the first ten numbers you count. Use the sets of fingers to help you.

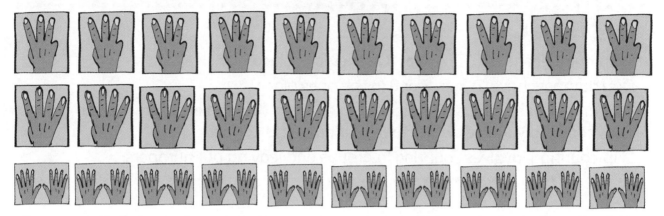

a Count by threes. Start at 0.

b Count back by threes. Start at 30.

c Count by fours. Start at 0.

d Count back by fours. Start at 40.

e Count by fives. Start at 20.

f Count back by fives. Start at 80.

g Count by tens. Start at 100.

h Count back by tens. Start at 100.

5 Look at the number line. What counting pattern does it show?

0   50   100   150   200   250   300   350   400   450

a Count by 50s to 100. How many 50s do you count?

b Count by 50s to 600. How many 50s do you count?

c Count by 100s to 600. Write the numbers that you count.

d Compare the pattern of skip-counting by 50 with the pattern for 5s and 10s. What do you notice?

Looking Back

A swimming pool is 50 metres long.
Sheilah swims 5 lengths of the pool. Jordan swims 8 lengths of the pool. Jessica cannot swim very well. She only makes half a length. Work out how far each girl swam.

Topic Review

What Did You Learn?

- Even numbers have 0, 2, 4, 6 or 8 in the ones place.
- Odd numbers have 1, 3, 5, 7 or 9 in the ones place.
- Odd and even numbers obey certain rules when you add or subtract them. The rules can help you decide whether your answer is correct.
- Skip-counting involves counting in groups and leaving out numbers.

Talking Mathematics

1 Sam wants to add 429 + 517. He knows the answer is either 945 or 946. How can he decide which answer is correct without doing any calculations?

2 Make up your own skip-counting sequence. You can start at any number.
   Tell your partner how to count. Check that they count correctly.

Quick Check

1 Here are six number cards.

| 14 | 22 | 19 | 17 | 31 | 34 |

a Which numbers are odd?

b Use the numbers on the cards to write three addition sums that will give an even number as the answer.

c Use the numbers on the cards to write three addition sums that will give an odd number as the answer.

d Will 34 − 19 give an odd or even number as the answer? Do the subtraction and check.

2 One number is incorrect in each of these counting patterns. Find it and say why it is incorrect.

a 33,   36,   39,   41,   45,   48

b 40,   36,   32,   28,   25,   20

c 400,   450,   500,   550,   650

d 138,   128,   118,   108,   100

3 Ten track stars each run 50 m for charity. How many metres is this altogether?

4 There are 15 children in a church group. They each pay $4.00 for an outing. How much money is collected in total?

5 How many sides are there on 18 triangles?

Topic 11 Money Workbook page 33

You can keep the change.

Key Words
adding
subtracting
difference
change
decimal point
counting on

▲ What does 'keep the change' mean? How much should the passenger pay? Do you think the taxi driver will be happy?

When we go shopping, we often do mathematical calculations to work out if we have enough money to buy something. In this topic, you are going to learn more about **adding** and **subtracting** money. You will learn how to work out the total cost of items by **counting on** in your head and you will learn to work out how much **change** you should receive when you pay for items.

Getting Started

1 You buy three cold drinks costing $5.00 each and a packet of chips for $2.00. What is the quickest way of adding up these numbers?

2 What could you buy with $10.00?

3 You have a $20.00 bill. You buy a T-shirt for $14.00, a magazine for $3.00 and some snacks for $1.50.

Would you get any change?

Unit 1 Working with Money

Let's Think …

You buy an ice cream that costs $1.80 and you give the ice-cream seller $2.00.

- How much money do you get back?
- How did you work that out? What calculation did you do? Did you work it out in your head?
- Why is it useful to be able to do calculations like this?

---

*When you have to **add** numbers, start with the biggest number and **count on**. You can count on quickly in 1s, 2s, 5s, 10s or 50s; for example:*

$13.00 + $3.00 = 13 + 1 + 1 + 1 = $16.00
$15.00 + $8.00 = 15 + 2 + 2 + 2 + 2 = $23.00

Change *is the* **difference** *between what an item costs and the money you give to the clerk when you pay for an item. To work out the change, you* **subtract**.

---

1 How much did they pay? Look at the picture and try to work out the answers without writing anything down.

AIR TIME/ PHONE CARDS $10.00
T-SHIRTS $16.00
CONES $6.00
BEACH CAPS $8.00
BEACH SANDALS $12.00
SWEETS $0.50
BOTTLES OF WATER $1.00

a Mario bought one T-shirt and an ice cream.

b Talise bought a beach cap and a pair of sandals.

c Brad bought two bottles of water and six sweets.

d Gabrielle bought air time for her phone and an ice cream.

2 Look at the picture again. What would you buy with each amount?

    a $5.00   **b** $10.00   **c** $15.00   **d** $20.00

3 Work out the change quickly in your head.

    a You buy an ice cream that costs $6.50. You give the clerk a $10.00 bill.

    b You buy some sweets that cost $1.50. You give the clerk two $1.00 bills.

    c You buy a cap that costs $7.25. You give the clerk a $5.00 dollar bill and three $1.00 bills.

*Remember to put the **decimal point** in the correct place when you add and subtract money.*

$5.00 is not the same as $500 or $0.50!

4 You have $10.00. Write down how much change you would get if you bought each of these items.

    a    **b**    **c**    **d**

5 Write down two problems for your partner to solve. Start like this.

    a You buy a ___ which costs ___ and a ___

        which costs ___. How much do you have to pay?

    b You buy a ___ and a ___. The ___ costs ___ and

        the ___ costs ___. You give the clerk ___. How much change will you get?

Looking Back

a Explain what it means to get or to give someone 'change'.

b What do you need to do to work out change: divide or subtract?

Topic Review

What Did You Learn?

- We use simple calculations when we go shopping.
- To work out the total cost of several items, we use addition.
- To work out change for a single item, we use subtraction.
- To work out the change when we buy several items, we first use addition and then subtraction.

Talking Mathematics

What is the mathematical word for each of these?

- to start with the biggest number and add the smaller numbers
- the difference between the cost of an item and the money you give to pay for the item

Quick Check

1 Calculate the change. Show your working out.

  a A book costs $7.25. You give the clerk $10.00.

  b A ball costs $11.60. You give the clerk $12.00.

  c A chocolate costs $2.90. You give the clerk $5.90.

  d A bag of fries costs $3.20. You give the clerk $4.50.

2 Use skip-counting to find the total amount of money. Write the amounts correctly using the $ or ¢ symbol and the decimal point.

Topic 12 Addition Strategies

Workbook pages 34–36

$345 + 79$ Est. $300 + 100 = 400$

$345 + 79$
→ $5 + 9 = 14 = 10 + 4$
→ $40 + 70 = 110 = 100 + 10$
→ $300 = 300$

$400 + 20 + 4 = \underline{424}$

```
 h   t   o
 3①  4①  5
+    7   9
─────────
 4   2   4
```

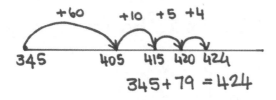

$345 + 79 = 424$

Key Words
add
addition
expanded
notation
place value
regrouping

▲ A teacher gave her class a sum. This is how three students worked out the answer. Talk about how each student worked. Which way looks easiest? Why?

You have done lots of **addition** sums before. Now you are going to work with bigger numbers. You will use addition facts, patterns, skip-counting and **place value** to find new ways of adding larger numbers.

Getting Started

1 Work out the missing number in each of these sums.

   a $50 + \square = 100$ b $20 + \square = 100$ c $\square + 70 = 100$

   d $\square + 60 = 100$ e $35 + \square = 100$ f $76 + \square = 100$

2 Tell your partner how you worked out the answers.

3 Make up six different addition sentences. Each one must have a total of 1 000.

Unit 1 Methods of Adding

Let's Think ...

- Do you remember what these words mean?

| Sum | Total | Add | Plus | In all | Altogether |

- Choose words from the box to complete these addition problems. Use each word only once.

  a ___ 17 and 23.

  b What is the ___ when you combine 120 and 30?

  c Nick has 32 blue marbles and 23 red ones. How many does he have ___?

  d Shayleen has 40 round counters and 45 oval counters. How many is this ___?

  e What is 90 ___ 120?

  f Find the ___ of 65 and 300.

- Now work out the answers to the problems.

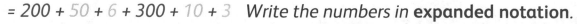

There are different methods of working out the answers to **addition** problems.

Breaking Down Numbers Using **Place Value**

256 + 313

= 200 + 50 + 6 + 300 + 10 + 3   Write the numbers in **expanded notation.**

= 200 + 300 + 50 + 10 + 6 + 3   Rewrite so the same places are next to each other.

= 500 + 60 + 9   **Add** the hundreds, *tens* and *ones.*

= 569   Write the number.

➡

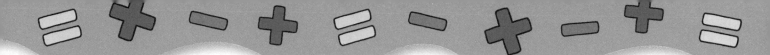

You can also work like this:

256 + 313

Add the ones:	6 + 3 = 9
Add the tens:	50 + 10 = 60
Add the hundreds:	200 + 300 = 500
Find the total:	500 + 60 + 9 = 569

Skip Counting in Chunks (With or Without a Number Line)

256 + 313   313 = 300 + 10 + 3

256 + 300 = 556

556 + 10 = 566

566 + 3 = 569

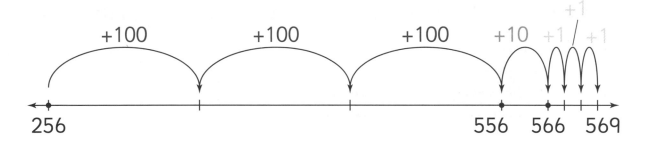

Regrouping

You can **regroup** numbers in any place to make it easier to add them. Regrouping works with any method.

255 + 189

Add the ones:	9 + 5 = 14	Regroup to 10 + 4
Add the tens:	80 + 50 = 130	Regroup to 100 + 30
Add the hundreds:	200 + 100 = 300	
Find the total:	300 + 100 + 30 + 10 + 4	
	= 400 + 40 + 4	
	= 444	

1 Break these numbers down using place value to add them. Regroup numbers if you need to.

a 134 + 456 b 544 + 328 c 231 + 654 d 129 + 432

e 409 + 532 f 124 + 303 g 456 + 122 h 412 + 247

2 Skip-count in chunks to add these numbers. Use a number line if you prefer.

a 430 + 136 b 135 + 332 c 324 + 406 d 198 + 430

e 809 + 130 f 735 + 206 g 436 + 122 h 546 + 271

3 Use the method you prefer to do these additions.

a 39 + 208 b 145 + 299 c 393 + 551 d 208 + 196

e 165 + 147 f 99 + 408 g 430 + 670 h 401 + 399

> *When you are solving word problems, you should always estimate before you calculate. This helps you to decide whether your answer is reasonable.*

4 Two cruise ships were docked in Nassau.
514 people got off the first ship and 328 got off the second ship.
How many people got off in all?

5 A large resort sold 312 tickets to the water park on Friday and 629 tickets on Saturday.

a How many tickets did they sell altogether on these two days?

b On Sunday, they sold double the number of tickets they sold on Friday. How many did they sell on Sunday?

c How many tickets did they sell over the three days?

Looking Back

a Make up three addition problems of your own.

b Swap with a partner and work out each other's answers.

Unit 2 More Adding

Let's Think ...

Shaquille wants to add 1742 and 325.

- How could he do this?
- Write down your ideas.

The methods you learned in Unit 1 can also be used to add numbers in the thousands.

Look at these examples carefully.

2 345 + 1 342 = ☐

2 + 5 = 7

40 + 40 = 80

300 + 300 = 600

2 000 + 1 000 = 3 000

3 000 + 600 + 80 + 7 = 3 687

1 876 + 4 159 = ☐

= 1 000 + 800 + 70 + 6 + 4 000 + 100 + 50 + 9

= 1 000 + 4 000 + 800 + 100 + 70 + 50 + 6 + 9

= 5 000 + 900 + 120 + 15

= 5 000 + 900 + 100 + 20 + 10 + 5

= 5 000 + 1 000 + 30 + 5

= 6 000 + 35

= 6 035

2 045 + 1 832 = ☐

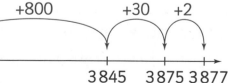

2045 + 1832 = 3877

1 Add. Use the method you find easiest, but show all your working.

　a. 1430 + 3098　　　　b 3458 + 3500　　　　c 2087 + 904

　d 983 + 1865　　　　　e 1291 + 308　　　　　f 4023 + 5195

2 The diagram shows the distance in kilometres from Nassau to some other places.

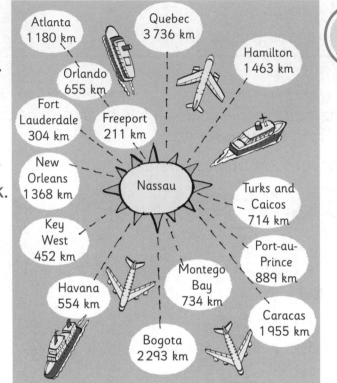

Atlanta 1180 km
Quebec 3736 km
Hamilton 1463 km
Orlando 655 km
Fort Lauderdale 304 km
Freeport 211 km
New Orleans 1368 km
Nassau
Turks and Caicos 714 km
Key West 452 km
Port-au-Prince 889 km
Havana 554 km
Montego Bay 734 km
Caracas 1955 km
Bogota 2293 km

　a How far is it to Freeport and back?

　b A plane flies from Port-au-Prince to Nassau and back. What is the distance it travels?

　c A cruise ship sails from Orlando to Nassau and then from Nassau to New Orleans. How far is this journey?

　d In one week, a businesswoman flies from Nassau to Havana and back and then from Nassau to Caracas and back. How many kilometres did she fly?

　e A pilot flies from Nassau to Hamilton and back. How far is this?

　f Double the distance from Nassau to Quebec. Add the distance from Nassau to New Orleans to the result.

Looking Back

Use distances from the diagram.
a Find two different distances with a sum of 5199 km.
b Which distance gives 4586 km when you double it?
c How many pairs can you make with a total of less than 1000 km?

Topic Review

Talking Mathematics

- Work with a partner.
- Choose one method of adding a 3-digit number to a 4-digit number.
- Make up an example to show your method.
- Pretend you are a teacher. Take turns to explain to each other how to use the method you have chosen to add larger numbers.

Quick Check

1 Write down the year we are in. Add this to the year we were in last year.

2 Add.

   a 29 + 106   b 432 + 575

   c 1 782 + 324   d 3 039 + 152

   e 2 005 + 409   f 6 092 + 1 949

3 a Write these six numbers in ascending order.

    1 678   1 840   1 562

    1 230   1 947   1 609

   b Add the greatest number to the smallest number.

4 a Write the smallest and greatest number you can make using the digits 4, 6, 0 and 5.

   b Add the smallest and greatest number.

   c What is double the smallest number?

   d Add 1 408 to the greatest number.

5 The mail ship delivered 683 pieces of mail one week and 718 pieces the next week.

   a How many pieces of mail did it deliver in the two weeks.

   b The ship covers a distance of 1 028 km per week. What distance does it cover in two weeks?

6 Mrs Johnstone's sister lives in Montreal in Canada. It is 2 291 km from Nassau. How far would she have to travel to visit her sister and return home?

Topic 13 Subtraction Strategies

Workbook pages 37–39

1) 264 – 153
= 111

$$200 + 60 + 4$$
$$- 100 + 50 + 3$$
$$\overline{100 + 10 + 1}$$

2) 361 – 205
= 156

−1−1−1−1−1 −100 −100
156   161   361

10
1 – 3 = 8
80 – 60 = 10
70
200 – 100 = 100
118

3) 281 – 163
= 118

Key Words

subtract
subtraction
place value
rename
minus
difference
take away

▲ This is Mario's homework. Follow his working. What strategies did he use to subtract the numbers? How did he get enough ones to subtract in the third example?

Subtraction problems involve **taking** one amount **away** from another to find out what is left. You are going to use some of the same strategies you used for addition to **subtract** larger numbers. You will do calculations and create and solve word problems that involve subtraction.

Getting Started

1 58 + 42 = 100.   Use this fact to write two subtraction facts.

2 Use the number facts you already know to work out the missing numbers in these number sentences.

a 100 – 40 = ☐   b 100 – 90 = ☐   c 100 – ☐ = 0

d 100 – ☐ = 100   e 100 – 45 = ☐   f 100 – ☐ = 35

3 Tell your partner what facts you used to find the answers in question 2.

4 Is 1 000 – 500 = 500 – 1 000? Explain why or why not.

Unit 1 Methods of Subtracting

Let's Think …

Shawnae says that 145 − 27 = 188. Without working out the calculation, explain how you can tell that her answer must be wrong.

Subtraction involves finding the **difference** between two numbers.

The symbol − means **minus** or **subtract**.

There are different methods of working out the answers to **subtraction** problems.

Breaking Down Numbers Using **Place Value**

464 − 323

Subtract the *ones*:	*4 − 3 = 1*
Subtract the *tens*:	*60 − 20 = 40*
Subtract the hundreds:	*400 − 300 = 100*
Add the hundred, tens and ones you have left:	*100 + 40 + 1 = 141*

Renaming

If you write the numbers in columns using place value, it is easy to **rename** *ones, tens or hundreds when you need to. Look at this example carefully.*

454 − 375

$$
\begin{array}{ccc}
& 100 & \\
300 & 40 & 10 \\
\cancel{400} + \cancel{50} + \cancel{4} & & 14 \\
300 + 70 + 5 & & \\
\hline
0 \quad\quad 70 \quad\quad 9 & &
\end{array}
$$

Step 1: Take 10 from 50 to make 14 ones
14 − 5 = 9

Step 2: Take 100 from 400 to make 14 tens (140)
140 − 70 = 70

Step 3: Subtract the hundreds
300 − 300 = 0
0 + 70 + 9 = 79

➡

Skip-Counting Backwards in Chunks

Counting back and subtracting in chunks (with or without a number line) means that you do not have to rename numbers.

641 – 456

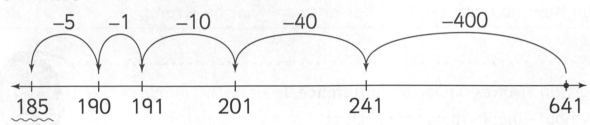

641 – 400 = 241
241 – 40 = 201
201 – 10 = 191
191 – 1 = 190
190 – 5 = 185

1 Expand the numbers using place value and then subtract.

a 98 – 43

b 48 – 23

c 43 – 29

d 432 – 141

e 866 – 251

f 487 – 302

2 Use the method you find easiest to do these subtractions.

a 562 – 228

b 205 – 108

c 773 – 329

d 465 – 184

e 327 – 106

f 893 – 145

g 628 – 325

h 409 – 299

i 907 – 488

Remember to estimate using rounding before you try to solve word problems. Compare your answer with your estimate to see if it is reasonable.

3 Two cruise ships were docked in Nassau. 514 people got off the first ship and 328 got off the second ship. How many more people got off the first ship?

4 A school library has 3 187 books. 450 books were taken out on loan. How many books were left in the library?

5 Use the words and phrases below to make up ten different subtraction word problems.

a Subtract
b Take away
c Difference
d How many remain
e How many more
f Minus
g Less
h How much further
i How much greater
j How much shorter

6 Exchange word problems with a partner. Use the method you prefer to work out the answers to your partner's problems. Show your working out.

Looking Back

Do these subtractions.

a 324 − 103
b 765 − 432
c 800 − 238

Unit 2 More Subtracting

Let's Think …

Trinity started with 2 564. She subtracted a number from it and got 1 235.

Now she cannot remember what she subtracted.

● How can you work out what the missing number is?

● Share your ideas with your group.

*You can use the same methods that you used in Unit 1 to **take away** numbers in the thousands.*

Read through these examples to make sure you understand how to do this.

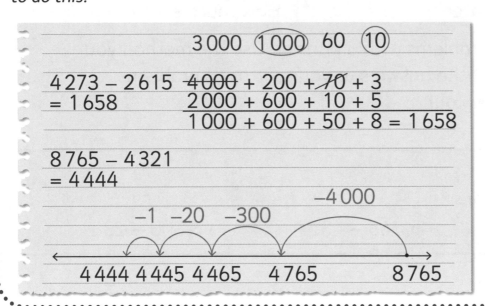

1 Subtract. Show your working.

a 3 425 – 1 245   b 4 009 – 2 389   c 5 000 – 2 040

d 4 624 – 3 654   e 8 472 – 2 573   f 8 050 – 4 099

g 1 234 – 987   h 7 534 – 4 128   i 3 045 – 2 199

2 The diagram shows the distance from Nassau to some other places.

a Which three places are furthest from Nassau? Write the three distances in order from closest to furthest away.

b Which is closer to Nassau: San Juan or Guatemala? How much closer is it?

c How much further is it to Vancouver than to Chicago?

d Nisha flew to Buenos Aires and Marie flew to Lagos. How much further did Marie fly?

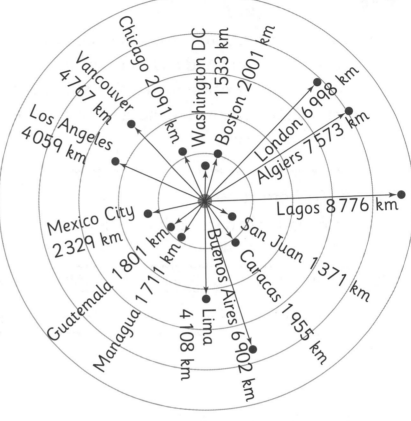

Chicago 2091 km
Vancouver 4767 km
Los Angeles 4059 km
Washington DC 1533 km
Boston 2001 km
London 6998 km
Algiers 7573 km
Lagos 8776 km
Mexico City 2329 km
San Juan 1371 km
Caracas 1955 km
Guatemala 1801 km
Managua 1711 km
Buenos Aires 6902 km
Lima 4108 km

e What is the difference between the distance to London and the distance to Mexico City?

f How many kilometres closer to Nassau is Caracas than Los Angeles?

g Subtract the shortest distance on the diagram from the longest distance.

Looking Back

Sharie wants to buy a car that costs $4527.00. She has $3649.00 saved. How much more does she need?

Topic Review

Talking Mathematics

Work in groups. Make a large chart with these headings. Write as many words as you can in each column.

Words that tell us to subtract	Words that tell us to add

Quick Check

1 The table shows how many emails a business received each day for two weeks.

	Monday	Tuesday	Wednesday	Thursday	Friday
Week 1	328	456	375	512	229
Week 2	194	521	556	1025	459

a Write the numbers for each week in ascending order.

b If Monday of the first week was on 4 July, what was the date on Tuesday of the second week?

c List all the numbers that have a 5 in the tens position.

d How many emails were received in total on Thursday and Friday of Week 1?

e What is the difference between the number of emails received on the first and second Thursdays?

f In the first week, how many more emails were received on Tuesday than on Monday?

g Subtract the smallest number of emails received in the two weeks from the highest number.

Topic 14 Graphs Workbook pages 40–43

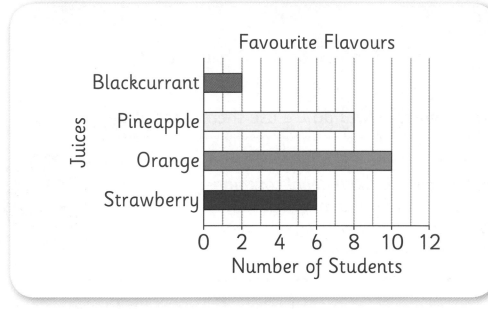

Favourite Flavours

Juices: Blackcurrant, Pineapple, Orange, Strawberry

Number of Students: 0 2 4 6 8 10 12

Workbook pages 40–43

Key Words
data
tally table
frequency table
horizontal bar graph
vertical bar graph
axis
title
labels
key
pictograph
scale
range
mode

▲ What information does this chart give us? How do you know this? Why do we display information in this way?

Graphs organize information in a way that is easy to see and read. You will remember that a **pictograph** uses symbols or pictures to show information. In this topic, you are going to learn more about showing information on graphs. You will also explore the graphs in order to understand all the information that they provide.

Getting Started

1 How can you collect information about what people like?

2 How can you record this information?

3 Can you predict what the results of a survey will be?

4 Why is it useful to collect data? Give an example.

Unit 1 Collecting and Recording Data

Let's Think ...

- Look at these tally marks. Which numbers do they show?

  a ⅏ I   b ⅏ ⅏   c ⅏ ⅏ IIII

- When do we use tally marks? Why do we use them instead of numbers?

Data is information that we collect about people or things.
*We use a **tally table** to record data using tally marks.*
*We use a **frequency table** to record data using numbers.*
Some tables have both tallies and numbers.

1 Work in groups. Do a survey of people's favourite sports.

a Make a tally sheet, like this, to record the data that you collect.

Sport	Tally
soccer	
cricket	
swimming	

b Ask 30 people which sport they like the most. Record the answers on your tally sheet.

c Make a frequency table using the data you have collected. Add up the tally marks and write the total for each sport; for example:

Favourite Sports	
Sport	**Number of People**
soccer	16

d In your groups, ask and answer questions about your table.

Looking Back

Work in groups. How can you make a chart to show how many people in your class have birthdays in each month of the year? What information do you need? How will you record the information? Once you have discussed this, make a chart.

Unit 2 Drawing Graphs

Let's Think ...

Look at these two bar graphs.

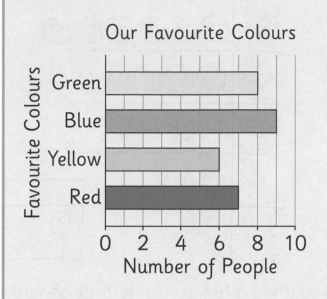

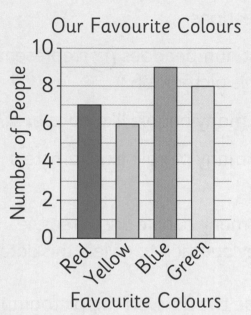

- Do they show the same information?
- How are the graphs different?
- Which graph do you think is clearer?
- Why do graphs need titles and labels?

In a **horizontal bar graph**, *the bars go across from left to right.*

In a **vertical bar graph**, *the bars go up from the bottom.*

All graphs have a vertical **axis**, *a horizontal axis, a* **title** *and* **labels**. *Some graphs also have a* **key**. *Bar graphs also have a* **scale** *which helps you read the numbers on the graph.*

1 Work in groups. Draw a vertical bar graph using the frequency table that you made about favourite sports. Your graph must have a title and labels (Number of People and Types of Sports).

You can draw the graph on page 41 of your Workbook.

> A **pictograph** *is a chart that gives information using pictures or symbols rather than numbers.*

2 Study this pictograph and answer the questions.

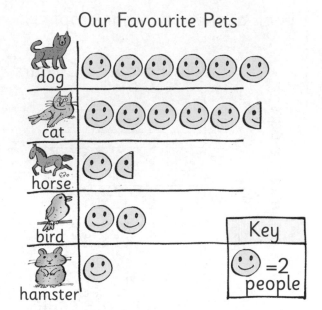

Our Favourite Pets

a What number does 😊 represent on this pictograph?

b How many people like dogs best?

c How many people like hamsters best?

d How many people answered the survey question to collect this data?

Key
😊 =2 people

3 Now use the data on the pictograph to draw a horizontal bar graph with the same information. You can do this in your Workbook on page 42.

4 Read this graph and answer the questions.

Our Favourite Ice-cream Flavours

a Is this a vertical or a horizontal bar graph?

b Which axis gives the numbers of people?

c What information does the other axis give?

d What information does this graph give us?

Looking Back
Name three different types of graphs.

Unit 3 Understanding Graphs

Let's Think …

● What is the range of ages in your class? What is the youngest age? What is the oldest age?

● What age are most students in your class?

*The **range** is the difference between the highest and lowest numbers in a set of data. You need to subtract the lowest number from the highest number to find the range.*

*The **mode** is the most frequently occurring value in a set of data.*

1 Arrange these values in order from the highest to the lowest. Calculate the range of the values in each set.

a 50, 23, 34, 93   b 19, 20, 11, 22, 101

2 There are 30 students in a class. The youngest is 7.

The oldest is 10. What is the range of ages in the class?

3 Look at the graph with the title: *Our Favourite Ice-cream Flavours*, on page 90. What is the mode of this set of data?

4 Work in groups. Carry out a survey in your class to find out which are the favourite subjects.

a Before you start, predict what you think the outcome of your survey will be. Which subjects do you think will be the favourites?

b Draw a graph to display your results.

Looking Back

1 What is the mode of a set of numbers?
2 What is the range of these numbers? 9, 15, 43, 21

Topic Review

Talking Mathematics

What is the mathematical word for each of these?

- a collection of information
- a bar graph in which the bars go across from left to right
- a chart that gives information using pictures or symbols instead of numbers
- a chart on which we record data using tally marks
- a chart on which we use numbers to record data

Quick Check

1. Write this tally in numbers: ||||| ||||| |
2. Why does a graph need to have a title?
3. How can you record the results of a survey?
4. Name two types of graphs. Draw an example of each and label them to show the important features.
5. Look at this graph. Do you think this graph is useful or not? Give a reason for your answer.

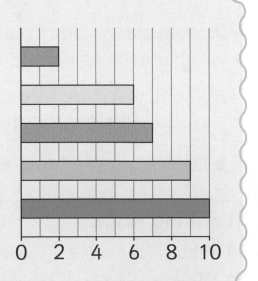

Topic 15 Problem Solving Strategies

Workbook pages 44–46

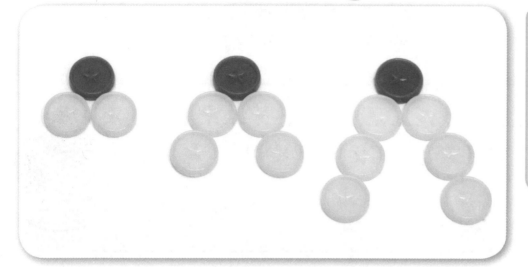

Key Words
choose
strategy
pattern
list
trial and error

▲ Asia made this pattern using plastic toys. What type of pattern is it? How many pieces will she need to build the next shape in her pattern?

You have already learned some **strategies** for solving problems and you have made up your own word problems in some topics. One way to get better at problem solving is to practice using different strategies so that you become comfortable with them. You can then **choose** the best **strategy** to use for particular types of problems.

Getting Started

How would you work to solve these problems? Talk about your ideas with your group.

1 Dwight is 9 years old. His grandma visited from the USA when he was born. She has visited once every other year since then. How many times has she visited?

2 Andrew is 3 years older than Lonnie. The sum of their ages is 19. How old is Andrew?

3 Zarea draws a rectangle on squared paper. Her rectangle covers 24 squares. Draw three different rectangles she could have drawn.

Unit 1 Choosing a Strategy

Let's Think ...

Mr and Mrs Winston have two children. Each child married and each couple had three children of their own. How many family members are there now?

Do you remember the steps you follow to solve problems?

Step 1: *Read the problem carefully.*

Step 2: List *the information that is given. Highlight the important words and numbers.*

Step 3: *Decide what you have to do: add, subtract, share. Draw a line under what you have to find out.*

Step 4: Choose a strategy. *You can draw a picture, use a* **pattern** *or write a number sentence. You could also use* **trial and error** *or make lists, charts or tables.*

Look at this example to see how drawing a diagram helped Ciara get the correct answer.

A square pen for goats is made with four posts on each side. How many posts are there?

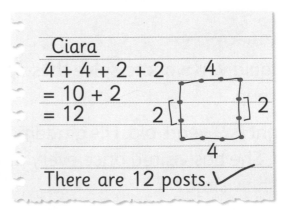

Jermaine
A square has 4 sides.
4 + 4 + 4 + 4 = 16
There are 16 posts. ✗

Ciara
4 + 4 + 2 + 2
= 10 + 2
= 12
There are 12 posts. ✓

Can you see why Jermaine's answer is incorrect?

1 Sami swims out 800 metres from the beach, then he floats for a while and the current takes him 200 metres back towards the beach. He then swims out another 600 metres. How far is he from the beach now?

  a Draw a diagram to show what is happening.

  b Write a number sentence for the problem and solve it.

2 Eight posts are used to make a square pen. There are an equal number of posts on each of the sides. How many posts are used for one side?

3 Ricardo cut a cake into 12 equal parts. If he ate $\frac{1}{4}$ of the cake, how many parts would be left?

4 Cybil fixes children's scooters and tricycles. Each scooter has one handlebar and two wheels. Each tricycle has one handlebar and three wheels. If there are 15 handle bars and 36 wheels in the workshop, how many scooters are there?

5 Look for a pattern and use it to help you solve the problems.

  a 1 + 2 + 3 + 4 + 1 + 2 + 3 + 4 + 1 + 2 + 3 + 4 + 1 + 2 + 3 + 4

  b 5 + 5 − 5 + 5 + 5 − 5 + 5 + 5 − 5 + 5 + 5 − 5

6 How many triangles are there in this diagram?

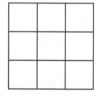

7 Janice says there are 10 squares in this diagram. James says there are 14 squares. Who is correct?

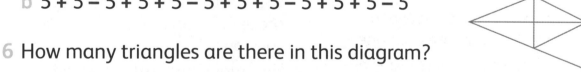

Looking Back

Dexter sells shells on the beach. Shells cost 10¢, 20¢, 30¢, 40¢, 50¢ or $1.00. He has $4.28 in his cash box. He sells three shells and his total increases to $5.98.

Which shells could he have sold? Write two possible combinations.

Topic Review

Talking Mathematics

Explain in your own words how to tackle a word problem. Use these phrases in your explanation:

- First I …
- Then …
- Next …
- After that …
- When I have an answer, I …

Quick Check

1 Sandra has these six number cards: **27 38 6 43 21 16**

  a Write the numbers in descending order.

  b What is the sum of the smallest and greatest numbers?

  c What is the difference between the smallest and greatest numbers?

  d Which of these numbers are even?

  e Which of these numbers would you count if you counted by 3s from 0?

2 Kendra thinks of a number.

  She doubles the number.
  Her number is now 28.
  She adds 12 to the result.

  What was the first number she thought of?

3 Paris made these three shapes with matchsticks.

  a Count the number of matchsticks she uses for each shape. What do you notice?

  b How many matchsticks would she need to build the next three shapes? Try to work this out without drawing or making the shapes.

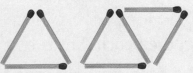

Shape 1   Shape 2   Shape 3

Topic 16 Measuring Length

Workbook pages 47–50

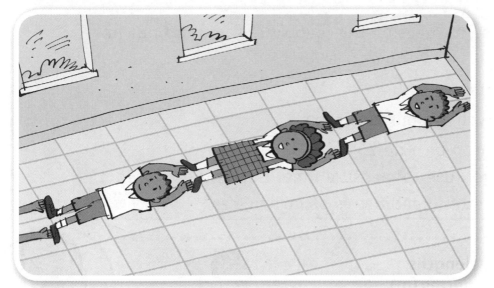

▲ What do you think these children are measuring? Is this a good way of measuring? How else could they measure?

Key Words
centimetre (cm)
decimetre (dm)
metre (m)
millimetre (mm)
length
height
width

We measure things all the time: for example, if you want to get a new bed, you will need to measure to see if it will fit in your bedroom. You already know that you can use your hands, feet and arms to make quick measurements. In this topic, you are going to learn more about the different units we can use to measure **length**, **height** and **width**.

Getting Started

1 Look around the classroom. Find two things that you could measure in centimetres (cm).

2 What could you measure with an open hand?

3 Would you measure the length of your little finger in metres? Why or why not?

4 Why would it be better to measure the length of a basketball court in metres rather than in centimetres?

Unit 1 Measuring in Metres

Let's Think …

● What could you use to measure the following if you did not have a measuring tape?
  ● the length of a rug
  ● the height of a wall in your classroom
  ● the width of the top of a desk
● What unit would you use to measure these things if you used a measuring tape: metres, centimetres or millimetres?

*We can measure longer **lengths**,*
***heights** and **widths** in **metres (m)**.*

Your armspan is about 1 metre (1 m).

1 a Estimate whether the following are more or less than a metre.

  ● the length of a poster in your classroom

  ● the length of a skipping rope

  ● the height of a door in your classroom

  ● the height of your partner

b Use a measuring tape, a metre stick or a piece of string that is one metre long to check your estimates.

c How accurate were your estimates?

2 Why would it be better to measure the length of your classroom in metres rather than in centimetres?

Looking Back

1 What is the abbreviation for metre: m or mm?
2 Name three things you can use to measure in metres.

Unit 2 Measuring in Millimetres, Centimetres and Decimetres

Let's Think ...

- How many line segments can you see in this drawing?
- Estimate the length of each segment.

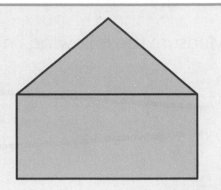

We can measure shorter lengths, heights and widths in **millimetres (mm)**, **decimetres (dm)** *or* **centimetres (cm)**.

The thickness of a nickel is about 1 mm.

Your handspan is about 10 centimetres (10 cm) or 1 decimetre (1 dm).

thumb

little finger

1 Work with a partner.

   a Make a drawing that has at least four line segments. One of the lines can be curved.

   b Discuss how you can measure each line segment.

   c Measure the segments.

2 Draw the following line segments in your exercise book and label them.

   a 4 cm   b 70 mm   c 1 dm

   d $8\frac{1}{2}$ cm   e 93 mm   f 27 mm

3 Kelly used her hands to estimate the size of a rug that would fit in her bedroom. She chose a new rug with her mom, but found it was too big for her room. How could she have made a better estimate?

4 Measure the following line segments in centimetres and millimetres.

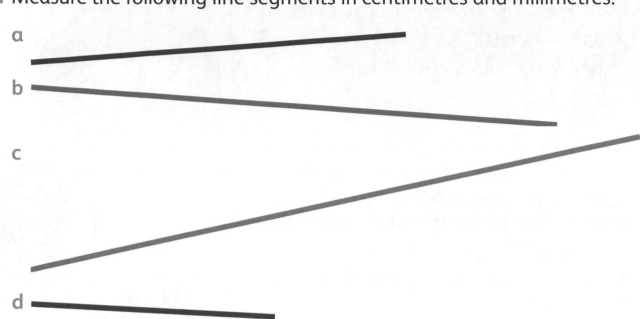

a

b

c

d

e

f

g

h

To measure a line that changes direction, like a path or a zigzag pattern, measure each part and then add up the measurements to get the total length.

5 Measure these paths in centimetres or in millimetres. Write the lengths in order from the longest to the shortest.

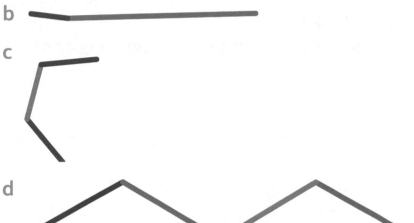

a

b

c

d

6 Use a ruler and different coloured pencils to draw each of these paths. Measure and write the total length under each path.

a 7 cm + 6 cm = ☐

b 70 mm + 6 mm = ☐

c 8 cm + 1 cm + 3 cm = ☐

d 2 cm + 2 cm + 2 cm = ☐

e 40 mm + 50 mm + 10 mm = ☐

f 30 mm + 80 mm + 30 mm = ☐

7 Draw a zigzag pattern or path of your own with a total length of 20 centimetres.

8 Wallace wants to share this strip of candy with two friends. He wants to measure very accurately to make sure everyone get exactly the same amount. What unit do you think he should use? How much do you think each person will get?

9 Which of these lines is the longest? Measure them and write the lengths in order from the longest to the shortest.

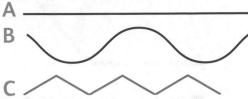

10 Work in groups. Why do you think it is important to have more than one way or unit to measure length, height and width?

Looking Back

1 Which unit would you use to measure each of these?

  a the length of a pencil

  b the length of your handspan

  c the height of your friend

  d the height of the trash can in your classroom

2 What do these line segments measure? Give the measurements in millimetres and in centimetres.

  a

  b

  c

  d

Topic Review

What Did You Learn?

- We use measurements in everyday life.
- We can measure length, height and width in metres (m), centimetres (cm), decimetres (dm) or millimetres (mm).
- We measure longer lengths in metres. We measure shorter lengths in centimetres, decimetres or millimetres.
- Your armspan is about 1 metre (1 m).
- Your handspan is about 10 centimetres (10 cm) or 1 decimetre (1 dm).

Talking Mathematics

What is the mathematical word for each of these?

- what we do when we guess how long something is
- what we do when we find out exactly how long something is
- a unit that is the length between your little finger and your thumb
- a unit that is the length between the tips of your fingers on each hand when you spread your arms open
- a unit which we can use to measure longer lengths and heights

Quick Check

1 What does each abbreviation stand for?

  a cm   b m   c dm   d mm

2 Which unit would you use to measure each of these?

  a the length of a finger   b the length of a tennis court

  c the thickness of a book   d the length of a pair of shorts

3 What do these line segments measure? Use your ruler. Write down the measurements in millimetres and in centimetres.

  a

  b

  c

  d

Topic 17 Multiplying and Dividing

Workbook pages 51–53

Key Words

multiplication
product
repeated addition
array
row
times
order
division
quotient
repeated subtraction
remainder

▲ How many rows of colour are there in this set? How can you use addition to find the total number of colours? How many equal groups of 2 are there? What happens if you make groups of 3?

In this topic, you are going to revise **repeated adding** and **subtracting** and learn a lot more about **multiplication** and **division**. You will make your own charts and use them to learn multiplication facts by memory. Once you know these, you will be able to do multiplication and division calculations with any numbers.

Getting Started

1 Why is it easier to count the eggs in the cartons than the ones in the basket?

2 Mrs Newton buys the blue carton of eggs. She uses 3 eggs every morning to make breakfast. How many days will the eggs last?

3 There are 19 eggs in the basket. How many groups of 2 can you make? How many eggs will be left?

4 How many eggs will you have if you buy 4 green cartons?

Unit 1 Understanding Multiplication

Let's Think ...

Hayley bought three packs of hair clips.
There are four clips in each pack.
How can you find the total number of hair clips?

Hair clips

*You can use **repeated addition** to find the total number of items in equal groups.*
4 + 4 + 4 = 12

*You can also think about this as a **multiplication**. There is 3 **times** 4.*
You can write this using a multiplication sign as 3 × 4 = 12.
*The answer to a multiplication calculation is called the **product**.*

*An **array** shows objects arranged in equal **rows**.*

Arrays are very useful for modelling multiplication. You can add the rows to find the total or you can multiply the number of rows by the number in each row to find the product.

This is an array for 3 times 4.

● ● ● ●
● ● ● ●
● ● ● ●

3 rows of 4
4 + 4 + 4 = 12
3 × 4 = 12

This array shows 4 times 3.

● ● ●
● ● ●
● ● ●
● ● ●

4 rows of 3
3 + 3 + 3 + 3 = 12
4 × 3 = 12

*The arrays show that you can multiply numbers in any **order** and still get the same product.*
3 × 4 = 12   4 × 3 = 12

Here is another example:

5 per row

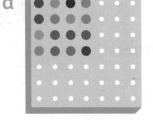

4 rows

Add 5 four times
5 + 5 + 5 + 5 = 20
4 × 5 = 20

number of rows

number of items in a row

1 A group of Grade 3 students made these arrays on pegboards. Write a repeated addition sum and a multiplication sentence for each one.

a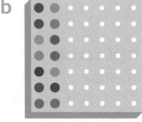

b

c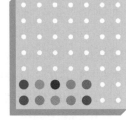

d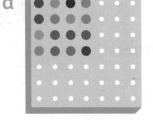

e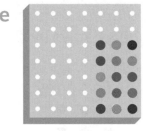

f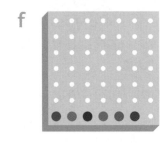

g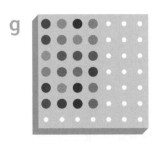

h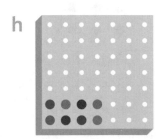

i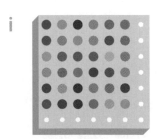

2 Draw an array to show each multiplication sentence. Copy the multiplication sentence and write the product.

a 3 × 2   b 5 × 3   c 4 × 2   d 8 × 1

3 Work in pairs. What is the missing number in each number sentence?
Explain how you know this.

a 4 × 8 = 32   b 5 × 3 = 15   c 9 × 2 = 18   d 6 × 5 = 30

8 × 4 = ☐   3 × 5 = ☐   2 × ☐ = 18   ☐ × 6 = 30

4 How many different arrays can you make with 12 counters?
Use counters to model them and then draw the arrays. Write
a multiplication sentence for each one.

5 Zarea made an array of counters. She wrote this number sentence using
her array: 3 × 9 = 27.

a How many rows were in her array?

b How many counters were in each row?

Looking Back

Look at the cupcakes in the box.
a How can you find the total by adding?
b What is the multiplication sentence for this arrangement?

Unit 2 Understanding Division

Division *is a way of sharing an amount equally or making equal groups. You can use division to find out how many equal groups there are in an amount.*

Zion has 20 beads. He uses 4 beads to make a bangle. How many bangles can be he make?

Beads 20 – 4 ⟶ 1 bangle

16 – 4 ⟶ 2 bangles

12 – 4 ⟶ 3 bangles

8 – 4 ⟶ 4 bangles

4 – 4 ⟶ 5 bangles
_____
0

Zion can make 5 groups of 4 with his 20 beads.

You can also use division to find out how many items there are in each group.

Kesia has 12 crayons. She wants to share them equally among 3 friends. How many crayons will each friend get?

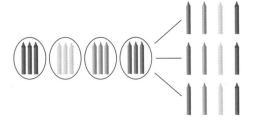

Here we have shared the crayons equally by putting one into each group until we have shared them all.

108

You can use **repeated subtraction** to solve a division problem. Repeated subtraction is another way of skip-counting backwards.

Zarea has 18 flowers. She puts 3 into a vase. How many vases will she need for all the flowers?

Start at 18. Subtract 3s until you reach 0. Count how many times you subtract to find the answer.

We have subtracted 6 times

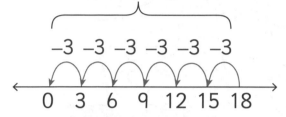

-3 -3 -3 -3 -3 -3

0   3   6   9   12   15   18

You can write this as a division sentence using a division sign. $18 \div 3 = 6$
The answer to a division problem is called the **quotient**.

1 This box of crayons contains 24 crayons.

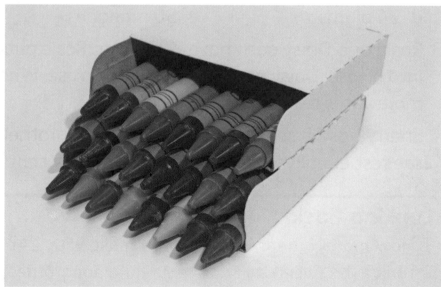

a Share the crayons equally among 3 children. How many will each child get?

b Make groups of 4 crayons. How many groups can you make?

c Share the crayons equally between 2 children. How many will each child get?

d What happens if you make groups of 5?

2 Nicky is helping her uncle pack fruit for his stall. The picture shows how many of each fruit she puts in a packet.

Work out how many packets Nicky would need to pack these numbers of fruits. Write a division sentence for each one.

a 15 avocados   b 16 pears   c 25 apples

d 24 plums   e 30 mangoes   f 21 bananas

3 Rosie and Daisy each have 24 flowers. Rosie puts 4 flowers into a vase. Daisy puts 3 flowers into a vase. Who will need more vases for their flowers? Why?

4 Sherry has 15 cents. She gives 3¢ to her brother then she shares the rest equally with her sister. How many cents does her sister get?

Looking Back

1 How many groups of 4 can you make with 24?

2 I have 18 sweets to share equally among 6 people. How many sweets will each person get?

3 Copy the number sentences. Fill in the quotient.

12 ÷ 2 = ☐   12 ÷ 3 = ☐

Unit 3 Multiplication and Division Facts

Let's Think ...

● Use the array of stars to work out these number facts.

  a 3 × 5   b 5 × 3   c 15 ÷ 3   d 15 ÷ 5

● What do you notice about this set of facts?

● What could we call this set of facts?

You need to learn your multiplication facts (times tables) by memory. Do not worry! If you use the patterns you already know from skip-counting and learn some new patterns and rules, you can master these facts very quickly: for example, you already know how to double all the numbers to 10, so you already know the 2 times table.

Facts for × 0

Think about an array that has rows with zero items in them. It does not matter how many rows of nothing you have, you will still have nothing.

Any number multiplied by 0 has a product of 0.

1 × 0 = 0

2 × 0 = 0

3 × 0 = 0

Facts for × 1

When you multiply a number by 1, the product is the number you started with.

1 × 1 = 1

2 × 1 = 2

3 × 1 = 3

Here are the times tables for 2, 3 and 5.

0 × 2 = 0	0 × 3 = 0	0 × 5 = 0
1 × 2 = 2	1 × 3 = 3	1 × 5 = 5
2 × 2 = 4	2 × 3 = 6	2 × 5 = 10
3 × 2 = 6	3 × 3 = 9	3 × 5 = 15
4 × 2 = 8	4 × 3 = 12	4 × 5 = 20
5 × 2 = 10	5 × 3 = 15	5 × 5 = 25
6 × 2 = 12	6 × 3 = 18	6 × 5 = 30
7 × 2 = 14	7 × 3 = 21	7 × 5 = 35
8 × 2 = 16	8 × 3 = 24	8 × 5 = 40
9 × 2 = 18	9 × 3 = 27	9 × 5 = 45
10 × 2 = 20	10 × 3 = 30	10 × 5 = 50

Multiplication facts are part of a fact family. Division is the inverse of multiplication. This means you can use one fact to find the others.

Look at this fact family.

3 × 6 = 18   6 × 3 = 18   18 ÷ 6 = 3   18 ÷ 3 = 6

1 This is a 'fact city' poster that a group of Grade 3 students made. Look at the blue building.

a How many windows on the first floor?

b How many windows on the first two floors?

c Use the blue building to find the product of 5 and 4.

2 You are going to work in pairs to make a 'multiplication fact city' that you can use to help you learn your times tables from 2 to 9.

You will need a large sheet of paper, crayons, magazines, scissors and glue.

Your teacher will tell you what facts you are to work with.

- Plan out your buildings carefully. You have to fit in 9 floors of windows and you must be able to fit enough windows on each floor.
- Cut out enough shapes for your 9 floors of windows.
- Paste them onto the building in neat rows.
- Colour and decorate your poster.

3 Use the fact buildings that your class has made to complete the multiplication facts on pages 51 and 52 of your Workbook.

Looking Back

Here are three multiplication facts.

2 × 7 = 14   6 × 7 = 42   9 × 4 = 36

Write the other three facts in each fact family.

Unit 4 Multiplying Bigger Numbers

Let's Think …

A bald eagle can carry four times its own body weight. If a female bald eagle weighs 14 pounds, what weight can she carry?

We can use place value and expanded notation to multiply bigger numbers.

Look at this example to see two different methods of working.

123 × 3 = 369

H	T	O			× 3
100	20	3		3	9
100	20	3		20	60
100	20	3		100	300
300 + 60 + 9					369

You may have to regroup ones and tens. Look at this example to see three different methods of working.

99 × 2 = 198

H	T	O			× 2		99
	90	9		9	18		× 2
	90	9		90	180		18 → 2 × 9
	①80 + ①8 regroup				198		180 → 2 × 90
100	10						198
100 + 90 + 8							

1 Multiply. Show your working out.

a 42 × 3   b 81 × 6   c 142 × 4

d 25 × 8   e 34 × 5   f 135 × 3

g 17 × 9   h 186 × 6   i 150 × 7

j 80 × 5   k 304 × 3   l 108 × 4

2 This is Jayne's homework. She did not show her working out. Check her answers. Did she get any wrong?

<u>16 June</u>

1. 3 × 37 = 81

2. 345 × 7 = 2413

3. 4 × 234 = 926

4. 402 × 3 = 1206

3 The island of North Bimini is 210 metres wide. A heron flies across the island 7 times. How far did it fly altogether?

4 The fishing limit for wahoo, dolphin and kingfish is 6 fish per person (any type). If 279 fisherman each catch this limit during one season, how many fish are caught in all?

5 Mrs Newton is buying plane tickets to visit her family. The tickets cost $85.00 each and she buys 4. How much will she pay?

6 How much would it cost to buy 8 concert tickets that cost $4.95 each? (*Hint*: remember $4.95 is the same as 495 cents.)

Looking Back

Multiply.

a 67 × 3   **b** 39 × 8   **c** 148 × 4   **d** 303 × 5

Unit 5 Dividing Larger Amounts

Let's Think …

Veronique has 68 stickers. She wants to stick the same number of stickers in each of her 4 sticker albums.

- How could she divide 68 by 4 to work out how many stickers to put in each album?
- Share your ideas with your group.

You can use multiplication facts to solve division problems.

$45 \div 5 = 9$ Think: $\boxed{} \times 5 = 45$

Look at this example. $48 \div 5$

You know that all multiples of 5 end in 5 or 0. So 5 cannot divide exactly into 48.

Think: what multiple of 5 is close to 48? $45 \div 5 = 9$ $48 - 45 = 3$

So, $48 \div 5 = 9$ with 3 left over.

The 3 that is left over is called a **remainder**. We can write the quotient as 9 r 3. For bigger numbers, you have to work with place value.

You can use a special division symbol to set out your work. We call this a division house.

$68 \div 4$ can be written with a division house like this:

You write the quotient above the division house.

the answer goes here

$4\overline{)68}$ ← this is a division house

Work through the example on the next page to see how to divide larger numbers. You still need to use your multiplication facts.

Step 1

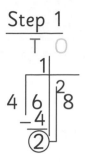

Think: How many times does 4 go into 6?

2 tens left over
20 + 8 = 28

Step 2

T O
1 | 7
4 | 6 ²8
−4
2

Think:

□ × 4 = 28

Step 3

Check by multiplying

4 × 17

× 4
7	28
10	40
	68 ✓

Because division is the inverse of multiplication, we can use multiplication to check the answer to a division.

42 ÷ 7 = 6   Check: 6 × 7 = 42 ✓

1 Use mental methods.

   a 36 ÷ 9   b 42 ÷ 8   c 72 ÷ 6   d 49 ÷ 5   e 57 ÷ 5   f 64 ÷ 8

2 a 72 ÷ 4   b 78 ÷ 6   c 92 ÷ 2   d 63 ÷ 3   e 52 ÷ 4   f 66 ÷ 6

3 A plane can hold 87 passengers. There are 3 seats in a row. How many rows are there?

4 A librarian has 76 books. She wants to share them equally between 4 shelves. How many books will she put on each shelf?

5 A baker made 90 buns. She packed them in boxes of 6. How many boxes did she use?

Looking Back

1 Complete these number sentences.

   a 45 ÷ 9 = □   b 49 ÷ 9 = □ r □

   c 20 ÷ 5 = □   d 24 ÷ 5 = □ r □

2 Work these out.

   a 4)‾24   b 4)‾48   c 4)‾55

Topic Review

What Did You Learn?

- Multiplication can be done as repeated addition or skip-counting.
- The times table for each number is a set of multiplication facts for that number.
- The answer to a multiplication is called the product.
- Multiplying by 0 always produces 0. Multiplying by 1 leaves the value of the number you are multiplying unchanged.
- You can use place value and expanded notation to multiply larger numbers.
- Division means splitting a number into equal groups or finding out how many items there are in each group.
- The answer to a division is called the quotient.
- You can use mental methods to divide.
- Some numbers do not divide exactly and you are left with a remainder.
- You can use a division house and place value to divide larger numbers.

Talking Mathematics

These statements each have an incorrect mathematical word in them. Find it and say what the correct word should be.

| The leftover number in a division is called a reminder. | The answer to a multiplication is the sum of the numbers. | You can use repeated subtraction to model multiplication. | When you combine equal groups to get a total, you are dividing. | When you put objects in equal rows, you call it a quotient. |

Quick Check

1 For each calculation, write the other three facts that are in the fact family.

  a 8 × 3 = 24   b 9 × 8 = 72

  c 63 ÷ 7 = 9   d 42 ÷ 6 = 7

2 What is the remainder when you divide 32 by 3?

3 A group of 20 children is arranged to make rows of 3. Draw an array to show this. How many children are left over?

Topic 18 Fractions Workbook pages 54–57

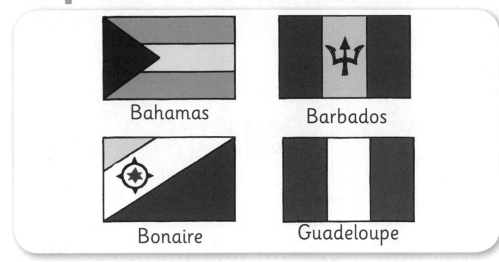

Bahamas

Barbados

Bonaire

Guadeloupe

Key Words
fraction
part
whole
numerator
denominator
equivalent
simplest form

▲ Look at the flags. Which flags are divided into equal parts?
Which flag is half blue? Which flag is one third blue?

Anything that is divided into equal **parts** can be described using **fractions**.
The flag of Barbados is divided into three equal parts. Each part is one
third of the flag. The flag of the Bahamas is not divided into equal parts,
so you cannot easily say what fraction of the flag is blue or yellow. In this
topic, you are going to represent fractions in different ways. You will also
compare and order them, find pairs of different fractions that are
equivalent and learn to how write fractions in their **simplest form**.

Getting Started

1 Look at the flags again. Use fractions to describe:

  a the part of the Guadeloupe flag that is red

  b the part of the Bonaire flag that is not blue

  c the part of the Barbados flag that is blue.

2 Draw a flag that is half red, one fourth white and one fourth blue.

3 Mario says that two fourths of the Jamaican flag are
green. Is he correct? Explain why or why not.

4 Find another flag that is divided into equal parts. Draw it, label the
fractions and write the name of country whose flag it is.

Unit 1 Understanding Fractions

Let's Think ...

The teacher asked her class to draw a diagram to show $\frac{1}{4}$.
Tonique and Asia drew these diagrams.

- How did each girl show the fraction?
- The teacher says they are both correct. Explain why.

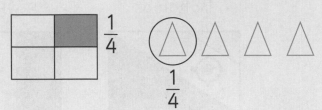

A **fraction** is an equal **part** of a **whole** shape or a set.

We use fractions symbols such as $\frac{1}{2}$ (one half) and $\frac{1}{4}$ (one fourth) to name the parts.

The fraction gives us information about the part or parts we are dealing with.

Look at this rectangle.

5 out of the 8 parts are shaded green.

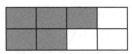

Fraction $\begin{cases} \frac{5}{8} \end{cases}$ numerator denominator

We say five eighths are green. We write this as $\frac{5}{8}$.

5 is the **numerator**. It tells us how many parts are green.

8 is the **denominator**. It tells us how many equal parts there are in the whole.

1 What fraction of each shape is blue?

a 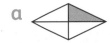 b  c  d

e  f   g 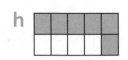 h

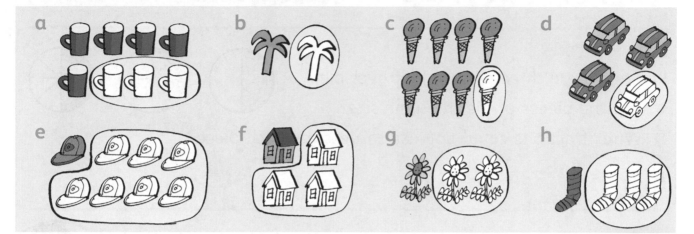

2 What fraction of each group is circled?

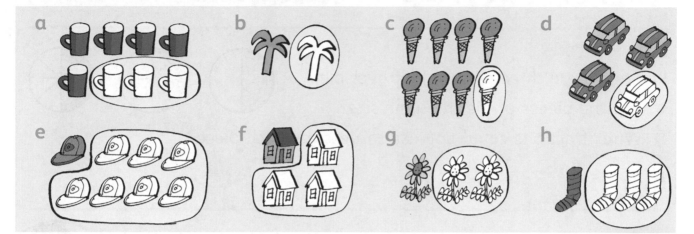

> *We can use >, = or < to compare fractions.*
>
> *When the fractions have the same denominator, we look at the numerator to decide which one is greater.*
>
> $\dfrac{3}{8}$ $\dfrac{5}{8}$ *Three parts is smaller than five parts, so $\dfrac{3}{8} < \dfrac{5}{8}$.*

3 Use **<**, **>** or **=** signs to compare the shaded fractions in each pair.

a

b

c

d

e

f

4 Read each statement. Say whether it is true or false. Correct any false statements.

a $\dfrac{3}{8} = \dfrac{4}{8}$ b $\dfrac{2}{5} > \dfrac{1}{5}$ c $\dfrac{8}{9} < \dfrac{9}{9}$ d $\dfrac{5}{6} > \dfrac{3}{6}$ e $\dfrac{2}{7} > \dfrac{5}{7}$

Looking Back

Write a fraction to describe each shaded part. Compare the shaded parts using **<**, **>** or **=**.

a   b   c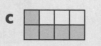

Unit 2 Equivalent Fractions

Let's Think ...

Look at these three pieces of leftover pie.

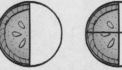

- Are the pieces of pie the same size?
- What fractions could you use to describe each piece?
- Is $\dfrac{1}{2} = \dfrac{4}{8}$?

We can write different fractions to describe the same part of a whole. Fractions that describe the same share of the whole are called **equivalent** fractions.

We can use fraction strips to compare fractions and find those that are equivalent.

Look at the fraction strips carefully.

Can you see that one half is the same length as two fourths? We write $\dfrac{1}{2} = \dfrac{2}{4}$.

Which other fractions on the fraction wall are equivalent to $\dfrac{1}{2}$?

1											

| $\frac{1}{2}$ | | | | | | $\frac{1}{2}$ | | | | | |

| $\frac{1}{3}$ | | | | $\frac{1}{3}$ | | | | $\frac{1}{3}$ | | | |

| $\frac{1}{4}$ | | | $\frac{1}{4}$ | | | $\frac{1}{4}$ | | | $\frac{1}{4}$ | | |

| $\frac{1}{5}$ | | $\frac{1}{5}$ | | $\frac{1}{5}$ | | $\frac{1}{5}$ | | $\frac{1}{5}$ | | |

| $\frac{1}{6}$ | | $\frac{1}{6}$ | | $\frac{1}{6}$ | | $\frac{1}{6}$ | | $\frac{1}{6}$ | | $\frac{1}{6}$ |

| $\frac{1}{7}$ | $\frac{1}{7}$ | $\frac{1}{7}$ | $\frac{1}{7}$ | $\frac{1}{7}$ | $\frac{1}{7}$ | $\frac{1}{7}$ |

| $\frac{1}{8}$ | $\frac{1}{8}$ | $\frac{1}{8}$ | $\frac{1}{8}$ | $\frac{1}{8}$ | $\frac{1}{8}$ | $\frac{1}{8}$ | $\frac{1}{8}$ |

| $\frac{1}{9}$ | $\frac{1}{9}$ | $\frac{1}{9}$ | $\frac{1}{9}$ | $\frac{1}{9}$ | $\frac{1}{9}$ | $\frac{1}{9}$ | $\frac{1}{9}$ | $\frac{1}{9}$ |

| $\frac{1}{10}$ | $\frac{1}{10}$ | $\frac{1}{10}$ | $\frac{1}{10}$ | $\frac{1}{10}$ | $\frac{1}{10}$ | $\frac{1}{10}$ | $\frac{1}{10}$ | $\frac{1}{10}$ | $\frac{1}{10}$ |

| $\frac{1}{11}$ | $\frac{1}{11}$ | $\frac{1}{11}$ | $\frac{1}{11}$ | $\frac{1}{11}$ | $\frac{1}{11}$ | $\frac{1}{11}$ | $\frac{1}{11}$ | $\frac{1}{11}$ | $\frac{1}{11}$ | $\frac{1}{11}$ |

| $\frac{1}{12}$ | $\frac{1}{12}$ | $\frac{1}{12}$ | $\frac{1}{12}$ | $\frac{1}{12}$ | $\frac{1}{12}$ | $\frac{1}{12}$ | $\frac{1}{12}$ | $\frac{1}{12}$ | $\frac{1}{12}$ | $\frac{1}{12}$ | $\frac{1}{12}$ |

1 For each fraction, write down an equivalent fraction.

a $\frac{1}{3}$ b $\frac{6}{9}$ c $\frac{1}{4}$ d $\frac{3}{6}$ e $\frac{2}{12}$ f $\frac{8}{12}$ g $\frac{1}{5}$

2 Fill in **<**, **=** or **>** to compare these fractions.

a $\frac{1}{8}$ ☐ $\frac{1}{4}$ b $\frac{1}{2}$ ☐ $\frac{6}{12}$ c $\frac{2}{5}$ ☐ $\frac{1}{3}$

d $\frac{2}{10}$ ☐ $\frac{2}{12}$ e $\frac{3}{6}$ ☐ $\frac{5}{10}$ f $\frac{4}{8}$ ☐ $\frac{3}{7}$

3 How many fractions can you write that are equivalent to 1 whole?

4 The picture shows what fraction of each container is filled with water.

a Is the green container more or less than half full?

b Which container is exactly half full?

c Which container has more water: the green or the blue?

d Write an equivalent fraction for each container.

e Order the fractions from least to greatest.

5 Draw a rectangular flag. Colour it so that half is blue, $\frac{1}{3}$ is green, $\frac{1}{6}$ is white and $\frac{2}{12}$ are yellow.

6 Use the fraction wall to help you write these fractions in order from greatest to smallest.

$\frac{3}{7}$ $\frac{1}{2}$ $\frac{6}{8}$ $\frac{2}{12}$ $\frac{9}{10}$ $\frac{3}{11}$

Looking Back

Arrange these fractions to make pairs of equivalent fractions.

$\frac{3}{12}$ $\frac{3}{4}$ $\frac{5}{10}$ $\frac{2}{3}$ $\frac{9}{12}$ $\frac{1}{2}$ $\frac{1}{4}$ $\frac{1}{3}$ $\frac{4}{12}$ $\frac{6}{9}$

Unit 3 More Equivalent Fractions

It is not always convenient to use fraction strips to find equivalent fractions but, if you understand how fractions work and apply some rules, you can multiply or divide to find equivalent fractions quickly and easily.

The first important rule is that when you × or ÷ any number by 1, it does not change its value.

$3 \times 1 = 3$ $487 \times 1 = 487$ $3 \div 1 = 3$ $487 \div 1 = 487$

The second rule is that 1 can be written as a fraction with the same numerator and denominator. (Look back to the fraction strip to see this.)

$1 = \frac{2}{2} = \frac{3}{3} = \frac{4}{4} = \frac{5}{5}$ and so on.

This means that if you multiply or divide the numerator and the denominator of a fraction by the same number, you are really multiplying or dividing by 1 and the value of the fraction does not change. In other words, you are finding an equivalent fraction.

Here are some examples.

$$\overset{\times 3}{\underset{\times 3}{\frac{1}{4}}} = \frac{3}{12} \qquad \overset{\times 7}{\underset{\times 7}{\frac{2}{5}}} = \frac{14}{35} \qquad \overset{\div 2}{\underset{\div 2}{\frac{10}{12}}} = \frac{5}{6} \qquad \overset{\div 7}{\underset{\div 7}{\frac{21}{28}}} = \frac{3}{4}$$

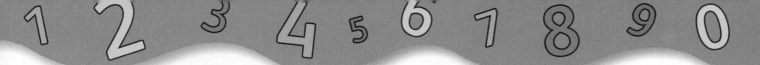

Fractions are in their **simplest form** when the only number that you can divide both the numerator and denominator by is 1.

$\frac{1}{2}, \frac{2}{3}, \frac{3}{4}$ and $\frac{4}{5}$ are all in their simplest form.

To find the simplest form of the fraction, divide the numerator and denominator by the same number until you cannot divide any more.

Here are two ways of simplifying $\frac{4}{16}$.

$$\frac{4}{16} \overset{\div 2}{\underset{\div 2}{\bigcirc}} = \frac{2}{8} \overset{\div 2}{\underset{\div 2}{\bigcirc}} = \frac{1}{4} \qquad \frac{4}{16} \overset{\div 4}{\underset{\div 4}{\bigcirc}} = \frac{1}{4}$$

If you use the biggest number that divides into both the numerator and denominator, you have to do fewer calculations.

1 Multiply each fraction by $\frac{3}{3}$ to find an equivalent fraction.

   a $\frac{1}{2}$ b $\frac{6}{9}$ c $\frac{7}{8}$ d $\frac{4}{7}$ e $\frac{11}{15}$

2 Divide each fraction by $\frac{5}{5}$ to find an equivalent fraction.

   a $\frac{15}{20}$ b $\frac{5}{10}$ c $\frac{35}{40}$ d $\frac{10}{15}$ e $\frac{20}{40}$

3 Multiply each fraction by $\frac{4}{4}$ to find an equivalent fraction.

   a $\frac{1}{2}$ b $\frac{2}{3}$ c $\frac{14}{15}$ d $\frac{21}{30}$ e $\frac{15}{20}$

4 Write each of these fractions in its simplest form. Show your working out.

   a $\frac{4}{12}$ b $\frac{5}{10}$ c $\frac{4}{8}$ d $\frac{10}{16}$ e $\frac{12}{16}$

   f $\frac{4}{6}$ g $\frac{8}{20}$ h $\frac{16}{20}$ i $\frac{15}{18}$ j $\frac{20}{25}$

5 Find the missing numerator or denominator in each pair of equivalent fractions.

a $\dfrac{2}{3} = \dfrac{\square}{6}$

b $\dfrac{1}{4} = \dfrac{5}{\square}$

c $\dfrac{3}{7} = \dfrac{9}{\square}$

d $\dfrac{3}{6} = \dfrac{\square}{18}$

e $\dfrac{3}{5} = \dfrac{6}{\square}$

f $\dfrac{1}{11} = \dfrac{\square}{22}$

g $\dfrac{10}{15} = \dfrac{2}{\square}$

h $\dfrac{6}{16} = \dfrac{3}{\square}$

i $\dfrac{12}{16} = \dfrac{\square}{4}$

6 Nikki spends 8 hours of every 24 hours asleep. What fraction of the day is she asleep? Give your answer in its simplest form.

7 16 of 180 children in a school were chosen to represent the district in a maths competition. Write this amount as a fraction in its simplest form.

8 There are 60 sweets in a bowl. Mr Smith ate 16, Mrs Smith ate 18 and the children ate 22.

a Who ate $\dfrac{3}{10}$ of the sweets?

b Write the fraction of sweets that are left in its simplest form.

Looking Back

Are these statements true or false?

a $1 = \dfrac{45}{45}$

b $3 \times \dfrac{3}{3} = 3$

c $\dfrac{12}{16} = \dfrac{4}{8}$

d $\dfrac{9}{10} = \dfrac{27}{30}$

e $\dfrac{35}{35} = \dfrac{100}{100}$

f $\dfrac{19}{20} = \dfrac{38}{40}$

g $\dfrac{4}{32} = \dfrac{1}{6}$

h $\dfrac{12}{16} = \dfrac{3}{4}$

Topic Review

Talking Mathematics

1 Sam got 1 sum correct, Lindy got 3 sums correct and Asia got 4 sums correct. How is it possible for them each to have half correct?

2 Michael says that you can simplify $\frac{5}{10}$ if you subtract 5 from 10. Is he correct?

How would you explain how to simplify $\frac{5}{10}$?

Quick Check

1 How many dots are there in $\frac{2}{3}$ of each group?

a 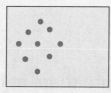   b    c 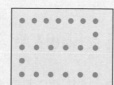   d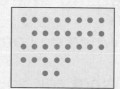

2 Arrange each set of fractions in ascending order.

a $\frac{1}{10}$, $\frac{1}{2}$, $\frac{1}{4}$, $\frac{1}{8}$ b $\frac{3}{6}$, $\frac{3}{8}$, $\frac{3}{10}$, $\frac{3}{12}$ c $\frac{1}{2}$, $\frac{4}{6}$, $\frac{4}{10}$, $\frac{9}{12}$

Topic 19 Working with Time

Workbook pages 58–61

▲ At what time was the first photograph taken?
Was the second photograph taken at the same time?
What is the difference?

We can use different units to measure time: for example, **seconds, days, weeks** or **years**. In this topic, you are going to learn to tell the time to the nearest five **minutes**. You will also practice changing times from one unit to another, and estimate and then calculate how long it takes to do things.

Getting Started

1 When do you finish school? At 3.00 a.m. or 3.00 p.m.?

2 What can you do in half an hour? How many minutes are there in half an hour?

3 Which is equal to 6 weeks: 42 days or 48 hours? How do you know?

Unit 1 Talking Time

Let's Think …

- How long is one minute? Count and stop when you think you have reached one minute. Did you count to 60?
- What can you do in 5 minutes?
- If you start a game at 11.00 a.m. and you play for two hours, do you finish in the morning or in the afternoon?

Hours, **minutes** *and* **seconds** *are units of time.*

1 minute = 60 seconds   *1 hour = 60 minutes*

$\frac{1}{2}$ *hour* **= 30 minutes** $\frac{1}{4}$ *hour* **= 15 minutes**

Watches and clocks have an **hour hand** *(the short hand) and a* **minute hand** *(the longer hand).*

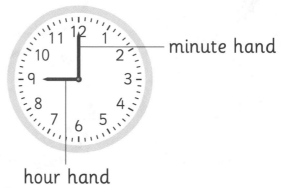

minute hand

hour hand

Some things take longer to do than other things.

1 Which of these activities can you do in ten minutes?

- have a shower
- read a whole book
- play a game of Bingo
- eat a small meal
- tell a story
- draw a picture

2 Choose a time from the box to match each of these.

a 2 minutes   b $\frac{1}{2}$ hour

c 1 minute   d $\frac{1}{2}$ minute

e $\frac{1}{4}$ hour   f 60 minutes

60 seconds	15 minutes
1 hour	120 seconds
30 minutes	30 seconds

3 Would these take place in the morning (a.m.) or in the afternoon or evening (p.m.)?

    a a birthday party starting at 3:00   **b** a trip to school on a bus

    c a trip home from school   **d** a sunrise at 5:30

You can work out the number of minutes past the hour on a clock by starting at 12 and counting on in fives until you get to the number that the minute hand is pointing to.

4 Say each time aloud. Then write down the time shown on each clock.

  a   **b**   **c**   **d**

  e   **f**   **g**   **h**

5 Kerrie gets on an aeroplane at 11:40 a.m. She is on the plane for 2 hours. At what time does the plane land?

6 You have 15 minutes to answer three questions in an exam. The questions are the same length and for the same number of marks. How much time should you spend answering each question?

Looking Back

1 Write down five things that you do every day. Write down the times that you do the activities. Write a.m. or p.m. after the times.

2 Complete the sentence.

    There are ＿＿ seconds in a minute and 60 ＿＿ in an hour.

Unit 2 How Long Does It Take?

Let's Think …

- Adderley started painting a picture at 5:00 p.m. He finished at 7:00 p.m. How long did it take him?
- Kamaya started knitting a scarf on Monday. She finished the scarf on Thursday. How long did it take her?

To work out how long something takes, you must know when it starts and when it ends. Then you can count on from the start time to the end time.

5 × 5 = 25 minutes

1 Four teams took part in a road race. They started and finished at different times. Read the table that shows these times and answer the questions.

Team	Starting Time	Finishing Time
Kamaya's team	10:05	11:30
Adderley's team	10:15	11:35
Sean's team	10:25	11:25
Zion's team	10:35	12:10

a Which team started earliest?

b Which team finished latest?

c Which team started at quarter past ten?

d Which team finished at twenty-five past eleven?

e How long did Adderley's team take to finish the race?

f Which team took the longest time?

g Which team took the shortest time?

h How many minutes were there between one team's starting time and the next team's starting time?

2 Choose the clock that shows the answer. Then write your answer using a.m. or p.m.

a You went to the beach at 11:20 a.m. You stayed there for 2 hours. At what time did you leave?

b Ernest went to his friend's house at 4:15 p.m. He stayed for three and a half hours. At what time did he leave?

c Torianne went to have her hair braided at 12:30 p.m. The braiding took two hours and twenty minutes. At what time was it finished?

d Mario fell asleep on his bed at 3:05 p.m. He woke up again 50 minutes later. At what time did he wake up?

3 Look at the calendar for July and answer the questions.

	JULY	

M	T	W	T	F	S	S
				1	2	3
4	5	6	7	8	9	10
11	12	13	14	15	16	17
18	19	20	21	22	23	24
25	26	27	28	29	30	31

a Mario's birthday is on the 16th. If today is the 7th, how many days until his birthday?

b You have basketball practice twice a week on Tuesdays and Fridays. How many practices do you have in July?

c Your teacher gives you a project on the 8th of July and says you must hand it in in two weeks' time. On what date must you hand it in?

d There is a sports tournament at your school. The tournament lasts 3 days. It starts on the 14th. On which day of the week does it end?

4 Work in pairs. Write two time problems for your partner to solve.

Looking Back

1 Write down the times or dates to answer these questions.

a You start reading a book at 3:30 p.m. You read for 2 hours. At what time do you finish?

b Mrs Williams makes a dress. It takes her 6 hours. She starts at 8:15 a.m. At what time does she finish?

c Ernest's birthday is in 5 days' time. Today is the 16th. What date is his birthday?

2 Write down these times in the shortest way possible.

a a quarter to seven in the morning

b thirty-five minutes past ten in the evening

c five to eight in the evening

Unit 3 Equivalent Times

Let's Think …
- Mario wants to know how many minutes there are in 3 hours.
  How can he work this out?
- Ashley thinks that there are five weeks in every month.
  Is he right or wrong?

To change **weeks** into days, multiply the number of weeks by 7.

To change **days** into hours, multiply the number of days by 24.

To change hours into minutes, multiply the number of hours by 60.

1 Change these times.

  a 3 weeks = ☐ days

  b 5 weeks = ☐ days

  c 4 days = ☐ hours

  d $2\frac{1}{2}$ days = ☐ hours

  e 6 hours = ☐ minutes

  f 9 hours = ☐ minutes

To change days into weeks, divide the number of days by 7.

To change hours into days, divide the number of hours by 24.

To change minutes into hours, divide the number of hours by 60.

2 Change these times.

  a 14 days = ☐ weeks

  b 28 days = ☐ weeks

  c 48 hours = ☐ days

  d 96 hours = ☐ days

  e 90 minutes = ☐ hours

  f 120 minutes = ☐ hours

3 Work out the answers to these riddles.

  a I am usually 365 days long, but sometimes I am 366 days long. What is my name?

  b I am 168 hours long. I am usually divided into days. What am I?

4 Work in groups. Make up three riddles about time to ask another group in the class.

5 A carnival lasted for 72 hours. It started on Thursday at 6:00 p.m. At what time and on which day did it end?

6 Ahkeem's brother started a course at university. The course started in September 2016 and it lasted for 22 months. How long was the course in years and months? When does his course end?

Looking Back

Complete the sentences.

a There are ☐ minutes in 1 hour.

b Each day has ☐ hours and each week has ☐ days.

c All months have at least ☐ days.

d There are usually ☐ days in a year.

e To change days into weeks, you ___ the number of days by 7.

f To change hours into ___, you multiply the hours by 60.

Topic Review

What Did You Learn?

- We can change units of time by dividing and multiplying numbers.

To Convert	Calculation
weeks into days	× the weeks by 7
days into hours	× the days by 24
hours into minutes	× the hours by 60
years to months	× the years by 12
days into weeks	÷ the days by 7
hours into days	÷ the hours by 24
minutes into hours	÷ the minutes by 60
months into years	÷ the months by 12

Talking Mathematics

What is the mathematical word for each of these?

- a time that lasts for 7 days
- a time that lasts for 60 minutes
- a time that lasts for 365 days
- a time that lasts for 4 to 5 weeks

Quick Check

Convert these times to the given units.

1 2 weeks = ☐ days

2 6 weeks = ☐ days

3 3 days = ☐ hours

4 $4\frac{1}{2}$ days = ☐ hours

5 12 hours = ☐ minutes

6 5 hours = ☐ minutes

7 21 days = ☐ weeks

8 35 days = ☐ weeks

9 72 hours = ☐ days

10 90 minutes = ☐ hours

11 730 days = ☐ years

12 48 months = ☐ years

Topic 20 Decimal Fractions

Workbook pages 62–63

0.25   0.50   0.60

Key Words
fraction
decimal
decimal point
tenths
hundredths
place value

▲ What does this photo show? What is the dot in each figure called? What does $0.50 mean?

Many of the numbers you see around you in daily life have a **decimal point**: for example, you may read that an athlete finished a race in 1.25 minutes or you might buy cola in a 1.5 litre bottle. You already know that you can write money amounts using the decimal point to show the cents as a **fraction** of one dollar. Now you are going to learn more about the decimal point and use it to write fractions (**tenths** and **hundredths**) in **decimal** form.

Getting Started

1 Look at the fraction strip carefully.

 a Is $\frac{1}{10}$ greater or smaller than 1?

 b Think about the place value chart. Where would you put a place for tenths? Why?

1 whole									
$\frac{1}{10}$	$\frac{1}{10}$	$\frac{1}{10}$	$\frac{1}{10}$	$\frac{1}{10}$	$\frac{1}{10}$	$\frac{1}{10}$	$\frac{1}{10}$	$\frac{1}{10}$	$\frac{1}{10}$

Unit 1 Tenths and Hundredths in Decimal Form

Let's Think ...

Noeleen has one dollar and twenty-three cents in her pocket.

● Which of these is the correct way of writing this?

$123 $12.30 $1.23 $0.123 $123.00

● Why are the others incorrect?

The diagram shows one whole divided into **tenths**.

Three tenths have been shaded.

We can write this as a **fraction** as $\frac{3}{10}$.

We can also write this as a **decimal** using the **decimal point** like this: 0.3.

$\frac{3}{10}$ is less than 1, so it has to be to the right of the ones place on the **place value** chart. The zero shows that there are no ones.

Hundreds	Tens	Ones	●	Tenths
		0	●	3

Look at this block. It has been divided into **hundredths**.

25 of the hundredths have been shaded.

We can write this as $\frac{25}{100}$.

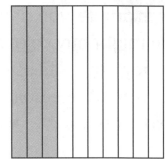

We can also write this as a decimal like this: 0.25.

This decimal has two places, so we need to extend the place value chart to include the hundredths.

Hundreds	Tens	Ones	●	Tenths	Hundredths
		0	●	2	5

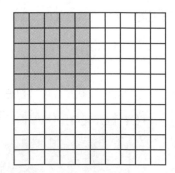

1 Write the fraction and the decimal for the shaded part of each square.

a 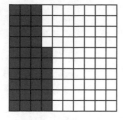   b 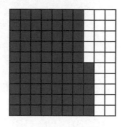   c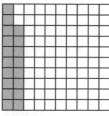

d 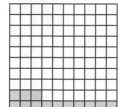   e 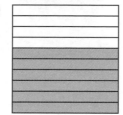   f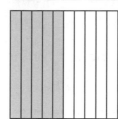

2 Write each fraction as a decimal.

a $\dfrac{9}{10}$ b $\dfrac{9}{100}$ c $\dfrac{99}{100}$ d $\dfrac{1}{2}$

3 There are 100 fish stickers on this sheet.

a Write the number of blue fish as a fraction with a denominator of 10 and as a decimal.

b What colour are 0.12 of the fish?

c Write a decimal to show what fraction of the fish are yellow.

d Which is greater: 0.2 or 0.25? Explain why.

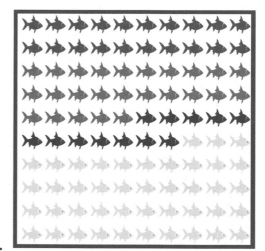

Looking Back

1 Write each of these as a decimal.

  a 9 tenths   **b** thirty-five hundredths   **c** $\dfrac{5}{100}$

2 Write the shaded part of each block as a decimal.

  a    b

Topic Review

What Did You Learn?

- Fractions such as $\frac{1}{10}$ and $\frac{1}{100}$ can be written in decimal form.
- A decimal is a number that uses place value and a decimal point to show fractions of a whole.
- $\frac{4}{10}$ can be written as 0.4. $\frac{25}{100}$ can be written as 0.25.

Talking Mathematics

- Write a number less than 1 which has 5 in the hundredths place and 7 in the tenths place.
- Explain how you would write 9 dollars as a decimal.

Quick Check

1 Express each of the following as a fraction and as a decimal.

a   b   c   d   e

2 Write four dollars and forty-five cents as a decimal.

3 Five students have written the decimal for $\frac{4}{10}$. Are their answers correct?

Tori	James	Asia	Brad	Charles
4.0	0.44	0.4	0.04	0.40

4 Is 0.08 smaller or greater than 0.1?

5 A sprinter beats her nearest rival by 0.92 seconds.

   a How would you say this number?   b What fraction of a second is this?

Topic 21 Capacity and Mass

Workbook pages 64–65

Key Words

measure

capacity

litres

millilitres

mass

kilograms

grams

balance scale

▲ Which container would you buy for storing water in the fridge? Which container would hold enough water for washing clothes? Which container would be heaviest if it was full of water?

We **measure capacity** and **mass** all the time: for example, when we make cookies, when we go to the doctor and when we go shopping. In this topic, you are going to investigate and compare units of measurement and think about which units to use for different measurements.

Getting Started

1 Which units of measurement are used to measure water and soda in bottles?

2 How can you measure rainfall?

3 Someone asks you, 'How much do you weigh?' What do they mean?

4 Which has a greater mass (is heavier): a balloon full of air or a packet of groceries?

Unit 1 Measuring Capacity

Let's Think …

- How much water do you drink every day?
- How do you know how much you drink?
- Do you bring water to school? How much water does your container hold?

The amount of liquid that a container holds is called its **capacity**.

We use **litres (L)** *and* **millilitres (mL)** *to* **measure** *how much liquid a container holds.*

We use millilitres to measure the capacity of small containers.

1 litre (L) = 1 000 millilitres (mL) 1 teaspoon holds about 5 mL.

1 a Make a list of containers that you use at home to hold liquids.

  b Draw each container and find out how much liquid it can hold.

  c Write the capacity in litres or millilitres under each drawing.

  d Compare your list with the lists of two other students in your class. Tell them what you use each container for.

2 Work in pairs. Estimate how many glasses of juice you can pour from each of these juice containers. Share your answers with the class.

3 Estimate the capacity of each of these. Choose the best answer.

a

b

c

d

100 mL or 100 L   220 mL or 220 L   1 mL or 1 L   500 mL or 50 L

4 Which capacity is greater? Copy and complete with < or >.

a 2 L ☐ 20 L   b 740 mL ☐ 1 000 mL   c 3 L ☐ 30 mL

d 2½ L ☐ 200 mL   e 100 mL ☐ 10 L   f ½ L ☐ 750 mL

5 Work in pairs. Discuss whether you would measure the following in litres or millilitres. Give reasons.

a tea in a teacup

b eye drops in a small medicine bottle

c yoghurt in a small tub

d gas in the tank of a car

e water in a bucket

f rain that falls in the garden

6 a Do a survey to find out how much water the students in your class drink in one day. Record your results on a chart.

b Do most of the students drink more or less than 1 L of water a day?

7 A basketball coach has three 2 L bottles of water for her team. There are 5 players in the team. The coach has cups with a capacity of 200 mL. How many cups of water can she give to each team member during the match? Write the number sentence and work out the answer.

Looking Back

Which unit would you use to measure each of these?

a a jug of juice

b a small amount of medicine

c a bucket of water

d a glass of water

Unit 2 Measuring Mass

Let's Think ...

Which is heavier?

- a book OR a pencil
- a bag of oranges OR a cake
- a pineapple OR a banana
- a mouse OR a hen

We measure **mass** in **grams (g)** and **kilograms (kg)**.

We use kilograms to measure heavy objects (such as bags of sand). We use grams to measure light objects (such as cookies).

1 kilogram = 1 000 grams

The apples on the **balance scale** have a mass of $1\frac{1}{2}$ kg.

$1\frac{1}{2}$ kg = 1 500 g

1 Would you use grams or kilograms to measure these masses?

a

b

c

d

e

f

2 Write down the mass shown on each scale.

a

b

c

d

3 Work in pairs. Collect five different objects. You can choose foods or other objects.

   a Make a chart like this or use the one on page 65 in your Workbook.

Object	Estimate of Mass	Actual Mass	Difference
1			
2			
3			
4			
5			

   b Estimate the mass of each object and fill in the chart.

   c Measure the actual mass of each object on a scale and record it on the chart.

   d Work out the difference between your estimate and the actual mass.

4 Ernest has a mass of $37\frac{1}{2}$ kg. Peter has a mass of 37 kg and 450 g. Who is heavier?

5 One shopping bag can hold 4 kilograms without breaking.

   a Write a number sentence to show the total mass of the groceries.

   b Work out how many shopping bags you need to carry these items.

Looking Back

1 Complete: **a** 1 L = ☐ mL   **b** 1 kg = ☐ g

2 Which unit would you use to measure each of these?

   a your own mass   **b** a cup of sugar   **c** a large pile of sand

Topic Review

What Did You Learn?

- The amount of liquid that a container holds is called its capacity.
- We use litres and millilitres to measure how much liquid a container holds.
- 1 litre (L) = 1 000 millilitres (mL)
- We measure mass in grams (g) and kilograms (kg).
- We use kilograms to measure heavy objects.
- We use grams to measure light objects.
- 1 kilogram (kg) = 1 000 grams (g)
- We can measure the mass of an object on a balance scale.

Talking Mathematics

What is the mathematical word for each of these?
- the amount of liquid that a container holds
- a unit we use to measure what a large container holds
- a unit we use to measure what a small container holds
- a unit we use to measure the mass of heavy objects
- a unit we use to measure the mass of light objects

Quick Check

1 How many millilitres in:
  a 1 litre
  b $2\frac{1}{2}$ litres?

2 How many grams in:
  a $\frac{1}{2}$ kilogram
  b 3 kilograms?

3 What would you measure each of these in? Choose the correct answer.
  a a small amount of sugar   grams / millilitres
  b a large bag of oranges   grams / kilograms
  c some flour to put in a cake   litres / grams
  d some oil to put in a cake   millilitres / grams

146

Topic 22 Exploring Shapes

Workbook pages 66–68

Key Words
line of symmetry
match
slide
flip
rotate
turn

▲ This famous building is in India. It is called the Taj Mahal. If you drew a line through the middle of the building, from top to bottom, what would you see? Are both sides of the building the same? Are both sides of the pond in front of the building the same?

We can see symmetry all around us. Some leaves and fruits are symmetrical. Buildings are often the same on both sides. Aeroplanes are also symmetrical. In this topic, you will find out more about symmetry. You will investigate how to find the **lines of symmetry** in different shapes and you will create symmetrical drawings and patterns.

Getting Started

1 Go outside and find two leaves or flowers or fruits. Would you have two equal halves if you cut any of these in half?

2 Is your face the same on both sides? Look in a mirror and check.

3 What will happen if you take these shapes and move them around or turn them over: a square, a pentagon, a triangle, an oval? Will the shapes change in any way?

Unit 1 Lines of Symmetry

Let's Think ...

Look at this photograph of a butterfly.

- Take a piece of string and use it to make a line through the middle of the butterfly, from the top to the bottom.
- Describe each side of the butterfly.
- What do you notice?
- Can you think of other animals that are like this?

Lines of symmetry *are lines that divide a figure in half.*

If you fold a shape in half, the fold line is the line of symmetry. The two halves of the shape are exactly the same. We say they **match**.

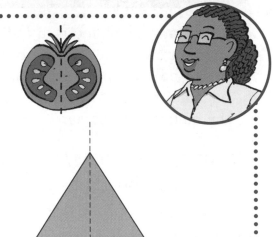

1 Cut out a square from a piece of paper. Fold the paper to make two matching parts.

2 Look at these shapes. How many different ways can you fold each one to get matching halves? Trace the shapes and try.

a    b   c    d

3 Fold a piece of paper in half. Draw a picture on the folded paper. Your picture must touch the fold. Cut out the picture and unfold it. What do you notice?

4 Work with a partner. Fold a piece of paper in half to get the line of symmetry. Draw half of a picture on the folded paper, along the fold. Open up the paper and let your partner complete the drawing by drawing exactly the same on the other half of the paper.

5 Collect pairs of objects that match; for example: shells, bottle tops, stones, sticks, and buttons. Make a symmetrical pattern with these objects.

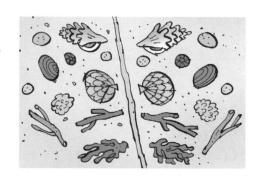

Looking Back

1 Complete this sentence.

If we fold a shape along the line of ___, we will get two shapes

that ___.

2 Can you see the symmetry in this photograph? Use a piece of string or a ruler to show where you would draw a line of symmetry.

Unit 2 Moving Shapes

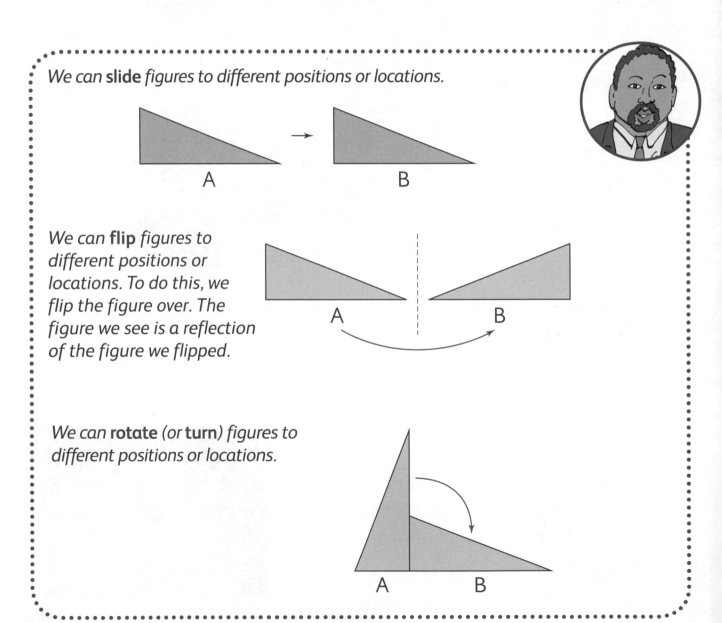

We can **slide** figures to different positions or locations.

A

B

We can **flip** figures to different positions or locations. To do this, we flip the figure over. The figure we see is a reflection of the figure we flipped.

A

B

We can **rotate** (or **turn**) figures to different positions or locations.

A

B

1 Take two different flat figures. One should be symmetrical and the other one not symmetrical. Slide them to different positions on your desk. Do they change?

2 Take the same figures you used in question 1 and flip them to reflect their shapes. What do you see now?

3 Take the same figures again and turn or rotate them. What has changed?

4 Look at the figures.

a Predict what will happen if we flip them. Will they look the same or different? Why?

b Choose one figure and make a drawing to show what the figure will look like if you rotate it to the right.

box

Looking Back

Look at the shape on the left.
a What has been done to the shape in A?
b What has been done to the shape in B?
c What has been done to the shape in C?

A   B   C

Topic Review

Talking Mathematics

What is the mathematical word for each of these?

- a shape with two halves that are exactly the same
- a line that we can draw through figures to divide them in half
- what we do when we turn a shape over
- the place where something is

Quick Check

1 Are these shapes symmetrical?

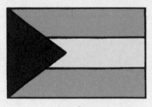 

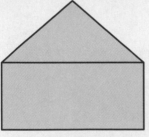  

2 Describe what happens if you fold a square exactly in half.
3 Draw what you will see if you flip this figure.

Topic 23 Perimeter and Area

Workbook pages 69–72

Key Words

perimeter
distance
side
closed plane shape
area
polygon
square units
square centimetre

▲ How could you work out if this TV set would fit on a wall in your house? What would you measure?

It is really useful to be able to measure **area** and **perimeter**: for example, if you want to find out how big a TV set is, you can measure around the set (the perimeter) or you can measure the surface of the screen (the area). In this topic, you are going to learn to measure the perimeter and area of different shapes.

Getting Started

1 What could you use to measure around the edges of a shape?

2 Look at a piece of string or ribbon that measures 30 cm. Is it long enough to measure around a book? Is it long enough to measure around the edges of a table in the classroom?

3 How many different rectangles can you make with 16 squares of paper? Try it and see.

Unit 1 Perimeter

Let's Think …

The perimeter of this cricket pitch is 4756 cm.

How could we work out this measurement?

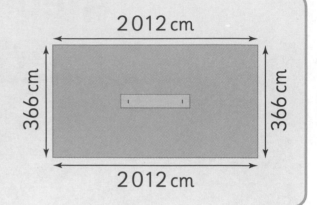

2012 cm

366 cm

366 cm

2012 cm

The **perimeter** *is the* **distance** *around the outside of a* **closed plane shape**.

A **polygon** *is a closed plane shape with straight sides. Regular polygons have sides of equal length. A square is a regular polygon.*

To calculate the perimeter of a shape, we can measure the length of each **side** *and then add the measurements to get the total.*

1 Measure the sides and work out the perimeter of these shapes.

a
b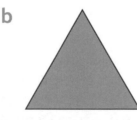
c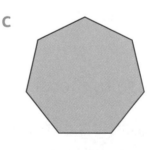
d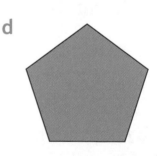

2 Draw a regular polygon with 4 sides. What do we call this polygon? Measure the sides and work out the perimeter. How else could you find the perimeter?

3 Discuss how you could work out the perimeter of these regular polygons by using multiplication.

4 Work in pairs. Each draw 2 different polygons and write down the perimeter of each one. Swap with your partner and ask them to check your measurements and calculations.

5 Draw polygons with the following perimeters.

   a 75 mm   **b** 80 mm   **c** 90 mm

   d 20 cm   **e** 24 cm

6 Take a walk around your local community with your teacher.

   a Find shapes that are polygons.

   b Measure the perimeter of one of the polygons.

   c Draw a diagram of the polygon you measured and label your diagram with the measurements.

Looking Back

1 Explain how to find the perimeter of a plane shape. Draw a diagram as part of your explanation.

2 Draw a polygon with a perimeter of 36 cm.

Unit 2 Area

Let's Think …

Look at these two shapes and the small squares next to the shapes.

- Would we need more small squares to cover shape A or shape B?
- Which shape could you cover with the small squares in the picture?

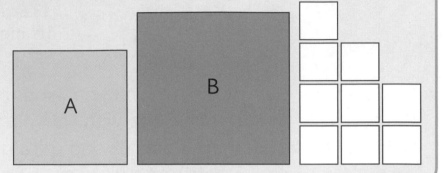

*The **area** of a figure is the number of **square units** that we need to use to cover the surface.*

*A **square centimetre** is a square with each side 1 cm long.*

The area of this rectangle is 8 square centimetres.

2 cm

4 cm

1 What is the area of each of these shapes? Give your answers in square centimetres.

a

b

c

2 Work in pairs. You will need two books of different sizes and square pieces of paper.

   a Estimate how many squares you will need to cover the first book.

   b Count the number of squares you think you will need.

   c Use the squares to cover the book.

   d Is the book covered? Do you have squares left over?

   e Now do the same with the other book.

3 Which of the animals in these drawings has the largest area?

4 Your teacher will give you some square pieces of paper. Use the squares to create a figure, for example an animal or a car. Then work out the area of your shape.

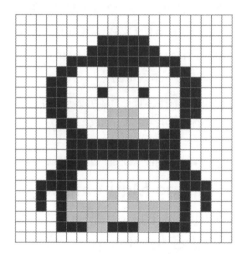

Looking Back

True or false?

a The area of a figure is the number of square units that we need to use to cover the surface.

b The area of this figure is 35 square centimetres.

c Two different shapes can have the same area.

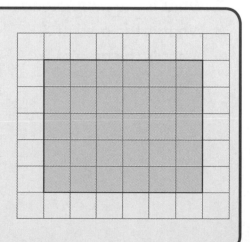

Topic Review

What Did You Learn?

- The perimeter is the distance around the outside of a closed plane shape.
- To calculate the perimeter, we measure the length of each side and then add the measurements.
- We can measure the perimeter in different ways, for example using tape measures, string, rulers or hand spans.
- A polygon is a closed plane shape with straight sides.
- Regular polygons have sides of equal length.
- The area of a figure is the number of square units that we need to use to cover the surface.
- Different shapes can have the same area because the area is arranged in different ways.

Talking Mathematics

What is the mathematical word for each of these?

- the distance around an object or closed plane shape
- the number of square units that cover a surface

Quick Check

1 What is the perimeter of A?
2 What is the perimeter of B?
3 What is the area of A?
4 What is the area of B?

A

B

Topic 24 Probability
Workbook pages 73–74

M	T	W	T	F	S	S
					1	2
3	4	5	6	7	8	9
10	11	12	13	14	15	16
17	18	19	20	21	22	23
24	25	26	27	28	29	30

Key Words
event
possible
impossible
certain
always
never
might

▲ This is the calendar for a month. Which month could it be? Is it possible that this is the calendar for February? Why or why not?

You already know that probability tells us whether **events** will **always** happen, **never** happen or sometimes happen. Now you are going to use the words that you learned last year to talk about probability and predict whether things will happen or not.

Getting Started

1 Look at the calendar again. If you tossed a counter onto the calendar, could it land on:

    a the 1st   b the 31st

    c the name of the month   d a Friday?

2 Why is it impossible for this to be the calendar for December?

3 Why is it impossible to decide exactly which month the calendar is for?

Unit 1 Certain, Possible and Impossible Events

Let's Think ...
Work in groups. Think about your school day. List:
- three things that will definitely happen today
- three things that might happen (but also might not happen)
- three things that will definitely not happen.

*Probability tells us how likely it is for something to happen or not happen. The thing we are talking about is called an **event**.*

*Some events are **certain**. This means they will **always** happen. If you roll a normal die, you are certain to get a number from 1 to 6.*

*Some events are **possible**. This means they **might** happen, but you cannot be certain that they will happen. If you roll a normal die, it is possible to get a 6 but you cannot be sure that you will get a 6.*

*Some events are **impossible**. They can **never** happen. If you roll a normal die, it is impossible for you to roll an 8.*

1 Say whether each event is possible, certain or impossible.

   a My pencil will break if I drop it.

   b The sun will rise tomorrow.

   c The bell will ring at the end of the school day.

   d Our next teacher will be a robot.

   e When I look at the clock, it will show 25 o'clock.

2 Write down three events that are certain for you today.

3 Look at these bags of counters. For each one, describe the probability of pulling out a blue counter. Use the words *certain*, *possible* or *impossible*.

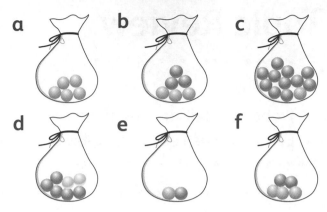

4 Colin has two 50¢ coins, one 10¢ coin and three 5¢ coins in his pocket. Decide whether each statement is true or false.

a It is possible for Colin to make $1.00.

b If Colin pulls out a coin, it is certain to have a value of 5¢ or more.

c It is possible for Colin to make $2.00 with the coins he has.

d It is impossible for Colin to pull 15 cents out of his pocket.

e If Colin pulls out a coin, he can be certain that it is Bahamian.

5 Look at these three spinners. For each one, write down an event that is certain, an event that is possible and an event that is impossible.

A   **B**   **C**

6 Draw a bag with five counters in it. Colour them five different colours so that it is possible to pull out a red or yellow counter from the bag but impossible to pull out a green or blue one.

Looking Back

Josiah is rolling a normal die with the numbers from 1 to 6 on it. Use the words *certain, possible* or *impossible* to describe each of these events.

a He will roll a 6 five times in a row.   **b** He will roll a zero.

c He will roll an odd number.   **d** He will roll a double.

e He will roll a number between 1 and 6.   **f** He will roll a number < 8.

Topic Review

Talking Mathematics

Use the words certain, possible or impossible to complete these sentences.

- It is ___ that it will rain today.

- It is ___ that our teacher will come to school by boat.

- It is ___ that we will have homework this week.

- The sun is ___ to set tonight.

- When you roll a die, it is ___ that you will get an odd number.

Quick Check

1 Draw a spinner with three colours so that it is:
  - possible to land on yellow
  - impossible to land on green
  - certain that you will land on blue, yellow or red.

2 What is the probability of each of these things?
  a It will rain and be sunny at the same time.
  b You will go to bed tonight.
  c There will be a flood at school today.
  d You will drive yourself to school tomorrow.

3 Zion made his own die using this pattern.

  a What events are possible if you roll Zion's die?
  b Draw a pattern of your own for a die. It should be possible to roll a 1, 2 or 5 on your die, but this should not be certain.
  c Compare your die with others in your group. Can you have different patterns for the dice, but have the same possible events?

Topic 25 Mixed Measures

Workbook pages 75–77

Key Words

Key Words
units
compound units
convert
equivalent
multiply
divide

▲ Which holds more liquid: the big carton or the five small cartons together? How do you know? How could you add these two amounts together?

We use different **units** of measurement to measure different lengths, masses and capacities. Some measures may be a combination of units: for example, someone may be 1 metre 45 centimetres tall. In this topic, you are going to change measurements from one unit to another and also add or subtract measurements with more than one unit.

Getting Started

1 How many centimetres are there in: 1 metre? 5 metres?

2 A piece of string is 700 centimetres long. How many metres is this?

3 What is the combined length of two pieces of string if each piece is 1 metre 30 centimetres long?

4 2 pints is the same amount as 1 quart. If I have 3 quarts, how many pint bottles can I fill?

5 Talk about how you worked out the answers in this activity. What operations did you use?

Unit 1 Working with Different Units

Let's Think ...

- What is the total length of the two ribbons?

  2 m 15 cm

- How much longer is the yellow ribbon than the green ribbon?

  1 m 10 cm

- What is the length of the yellow ribbon in centimetres?

A measurement that has two different **units** is called a compound measurement.

1 m 10 cm and 3 feet 4 inches are both compound measurements.

To add or subtract **compound units**, you can treat the units a bit like place values and work with each unit in turn.

For example:

1 m 45 cm + 2 m 38 cm

```
   m     cm
   1     45
+ 2     38
────────────
  3 m   83 cm
────────────
```

4 m 80 cm − 2 m 50 cm

```
   m     cm
   4     80
− 2     50
────────────
  2 m   30 cm
```

Add cm   45 + 38
          = 40 + 30 + 5 + 8
          = 83 cm

Subtract cm   80 − 50 = 30 cm

Subtract m   3 − 2 = 1 m

Add m 1 + 2 = 3

You must write the units in your answers.

You may have to rename units and **convert** from one unit to another to find an **equivalent** measurement when you do this.

> To convert from a big unit, such as metres, to a smaller unit, such as centimetres, you have to **multiply**. 1 m is the same as 100 cm, so you have to multiply by 100.
>
> 3 m = ☐ cm   3 × 100 = 300 cm   Think: There are 100 cm in 1 m.
>
> To convert from a small unit, such as mm, to a bigger unit, such as cm, you have to **divide**. There are 10 millimetres in 1 cm, so you divide by 10.
>
> 120 mm = ☐ cm   120 ÷ 10 = 12 cm   Think: There are 10 mm in 1 cm.

1 For each pair of ribbons, find the total length and the difference in length.

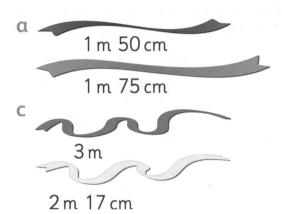

a
1 m 50 cm
1 m 75 cm

c
3 m
2 m 17 cm

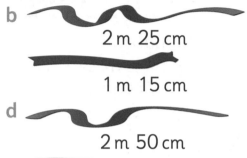

b
2 m 25 cm
1 m 15 cm

d
2 m 50 cm
98 cm

2 Convert each of these measurements to centimetres.

a 4 m   b 12 m   c 150 m   d $9\frac{1}{2}$ m   e 10 mm   f 120 mm

3 There are 3 feet in 1 yard. Mr Gleeson measures his path and finds it is 15 feet long. How many yards is this?

4 There are 12 inches in 1 foot. Shenade is 3 feet 4 inches tall.

a What is her height in inches?

b An inch is approximately $2\frac{1}{2}$ cm. Approximately how many millimetres are there in 3 inches?

Looking Back

What is:

a 1 m 48 cm + 3 m 98 cm   b 2 m 50 cm − 1 m 55 cm?

Topic Review

Talking Mathematics

Discuss the answers to these questions with your partner.

Which is bigger: 3 metres or 3 feet?

What would I need to know and do to convert a measurement in quarts to a measurement in gallons?

This measurement is in millimetres. I need it to be in centimetres. What do I do?

Quick Check

1 Convert each of these measurements to the units shown.

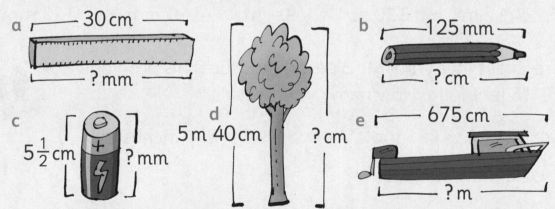

a 30 cm — ? mm

b 125 mm — ? cm

c $5\frac{1}{2}$ cm — ? mm

d 5 m 40 cm — ? cm

e 675 cm — ? m

2 For each pair of measurements, find the total length and the difference between the lengths. Show all your working out.

a 2 m 56 cm and 3 m 16 cm

b 5 m 41 cm and 7 m 81 cm

c 4 m 95 cm and 1 m 99 cm

d 8 m 63 cm and 6 m 83 cm

3 There are 4 quarts in a gallon. How many quarts of gas does a 20 gallon tank hold?

Topic 26 Looking Back

Revision A

1 If Zion counts from 250 to 650 by hundreds, which of these numbers will he count?

200, 350, 400, 500, 550, 600

2 Fill in the missing numbers.

a 4 350, 4 400, ☐ 4 500, ☐ ☐

b 9 150, 9 050, ☐ 8 850, ☐ ☐

3 Compare these numbers using the < or > symbols.

a 2 768 ☐ 3 529

b 3 394 ☐ 1 921

4 Round these numbers to the nearest ten or hundred and estimate the answer to each calculation. Then calculate, showing your working out.

a 54 + 76   b 130 − 85

c 3 190 − 2 623   d 4 218 + 3 373

5 What is the time on each clock.

a    b

6 A school principal drew this graph.

Type of Event	Number of Students
High jump	옷옷옷옷옷옷옷
Sprinting	옷옷옷옷옷옷옷옷
Javelin	옷옷옷옷옷
Marathon	옷옷옷옷옷옷옷옷옷옷

Key 옷 = 2 students

a What type of graph is this?

b Which event was most popular?

c How many students chose sprinting as their favourite event?

d Draw a bar graph to show this data.

Revision B

1 How many:

  a days in two weeks

  b centimetres in 3 metres

  c minutes in 3 hours

  d days in a leap year?

2 Write the correct mathematical word.

  a A flat shape that has length and width is a ___ shape.

  b A round solid that has no edges or faces is a ___.

  c A rectangle with all its sides equal is a ___.

  d A solid with six faces, all the same shape and size is a ___.

3 Solve each problem. Show your working.

  a 1 kilogram of bananas costs $3.00. How much will 5 kg cost?

  b A relay team has four runners who each run 100 metres. What is the total distance they run?

c There are 12 cartons of juice in a box. How many boxes would you need to give 18 children a carton of juice each?

d Four identical bags of sweets cost $5.00. What is the price of one bag?

4 What fraction of each shape is shaded?

a   b   c   d

5 In a group of 12 children, 8 have long hair. What fraction have short hair? Write the answer in its simplest form.

6 Look at this fraction: $\dfrac{3}{7}$

  a What is the numerator?

  b What is the denominator?

  c Is $\dfrac{3}{7}$ greater or smaller than $\dfrac{3}{8}$?

  d $\dfrac{3}{7} = \dfrac{9}{\square}$

7 I am thirteenth in a line of 20 people. How many people are:

  a in front of me?

  b behind me?

8 A nurse collected this data about families that visited the clinic.

Number of Children in Families	
1 child	HHT I
2 children	HHT HHT III
3 children	HHT
4 children	IIII
5 children	II

a What type of table is this?

b How many families had 2 children?

c How many children did 6 families have?

9 Measure length **a** in centimetres and length **b** in millimetres.

a

b ▬▬▬▬▬▬▬▬▬▬▬▬▬▬

10 Measure the sides of this rectangle and calculate its perimeter.

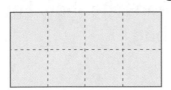

11 Each of the squares in this shape has sides 1 cm long.

a What is the perimeter of the shape?

b What is the area of the shape?

12 Write the length of each line segment as a compound unit using cm and mm. Calculate the combined length of the two line segments.

a

b

13 Name each shape. Write down three properties of each one.

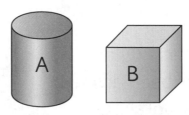

14 How has the white shape been moved to form the red shape in each of these pictures?

a   b   c

Revision C

1 What decimal fraction is shown by the shaded part of each square?

a 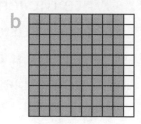 b

2 Write a multiplication and division fact family for each array.

a ★ ★ ★
  ★ ★ ★
  ★ ★ ★
  ★ ★ ★
  ★ ★ ★

b ★ ★ ★ ★ ★
  ★ ★ ★ ★ ★
  ★ ★ ★ ★ ★
  ★ ★ ★ ★ ★

c ★ ★ ★ ★
  ★ ★ ★ ★
  ★ ★ ★ ★
  ★ ★ ★ ★
  ★ ★ ★ ★

3 Calculate:

a 480 ÷ 8   b 43 ÷ 7

c 142 × 4   d 108 × 6

4 This cylinder is 48 cm high. It has been divided into four equal slices.

a How high is each slice?

b What is the combined height of three slices?

5 Ashanti has $64.00. This is four times as much as Joe. How much does Joe have?

6 Which of these shapes are symmetrical?

A   B   C   D

7 Complete the sentences about this spinner.

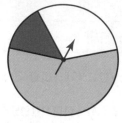

a It is possible for the spinner to land on ___.

b You can be certain that the spinner will land on ___.

c It is impossible for the spinner to land on ___.

8 A teacher is making bean bags to use in her classroom. She has four packets of beans. Each packet has a mass of 150 grams.

a What is the total mass of the beans?

b She wants to make 8 bean bags of equal mass. Work out what mass of beans she should put into each bag.

Key Word Reference List

The key words that you learned this year are listed here in alphabetical order. If you cannot remember the meaning of a word, turn to the page number that is given next to the word. Read the definition and look at the pictures or examples to help you remember what the word means.

a.m. (page 36)

add (pages 45 and 70)

addition (pages 44 and 74)

always (page 160)

approximate (page 60)

area (page 156)

array (page 105)

ascending (page 16)

axis (page 89)

backwards (page 66)

balance scale (page 144)

boil (page 29)

calendar (page 40)

capacity (page 42)

cardinal numbers (page 6)

Celsius (page 28)

centimetre (cm) (page 99)

certain (page 160)

change (page 70)

choose (page 94)

circle (page 52)

closed plane shape (page 154)

closed shape (page 51)

compare (page 23)

compound units (page 164)

cone (page 53)

corners (pages 51 and 53)

count (page 2)

counting (page 6)

counting on (page 70)

cube (page 53)

cylinder (page 53)

data (page 88)

day (pages 34, 38, 134)

decimal (page 138)

decimal point (pages 71 and 138)

decimetre (dm) (page 99)

denominator (page 120)

descending (page 16)

difference (pages 70 and 81)

digit (pages 2, 20, 58)

dimensions (page 53)

distance (page 154)

division (page 108)

edges (page 53)

element (page 14)

equivalent (page 122)

estimate (pages 30 and 60)

even (page 64)

event (page 160)

exact (page 60)

expanded form (page 20)

face (page 53)

fact (page 44)

fact family (page 46)

Fahrenheit (page 28)

flip (page 150)

forwards (page 66)

fraction (pages 120 and 138)

freeze (page 29)

frequency table (page 88)

grams (g) (page 144)

graph (page 3)

groups (page 66)

growing pattern (page 14)

height (page 98)

horizontal bar graph (page 89)

hour hand (page 129)

hours (pages 34, 38, 129)

hundreds (page 20)

hundredths (page 138)

impossible (page 160)

interval (page 66)

key (page 89)

kilograms (kg) (page 144)

labels (page 89)

length (page 98)

line (page 55)

line of symmetry (page 148)

line segment (page 55)

list (page 94)

litres (L) (page 142)

mass (page 144)

match (page 148)

measure (page 142)